Mobile and Wireless Networks

Advanced Networks Set

coordinated by
Guy Pujolle

Volume 2

Mobile and Wireless Networks

Khaldoun Al Agha
Guy Pujolle
Tara Ali-Yahiya

ISTE

WILEY

First published 2016 in Great Britain and the United States by ISTE Ltd and John Wiley & Sons, Inc.

ISTE Ltd
27-37 St George's Road
London SW19 4EU
UK

www.iste.co.uk

John Wiley & Sons, Inc.
111 River Street
Hoboken, NJ 07030
USA

www.wiley.com

Library of Congress Control Number: 2016943882

British Library Cataloguing-in-Publication Data
A CIP record for this book is available from the British Library
ISBN 978-1-84821-714-0

Contents

Preface

The world of mobile and wireless networks is only 10 years old, but is still expanding and evolving. Instead of settling in to a steady state, the changes are accelerating. The upcoming 3 to 5 years will expand on the current developments under 5G. This book aims to present the state-of-the-art in the field of mobile and wireless networks, and to anticipate the arrival of new standards and architectures.

After a description of the existing standards, mainly 2G, 3G and LTE, this book addresses LTE-A, which is the first 4G release, and provides a first indication of 5G as seen through the normalizing bodies.

4G technology is described in detail with the different LTE extensions related to the arrival of femtocells, the increase to 1 Gbps capacity, and relay techniques. 5G is also discussed to show what can be expected in the near future, and more precisely. A chapter is devoted to "small cells" that allow offloading techniques for discharging large antennas, and enable heterogeneous networks through integration with the normal macrocell.

Subsequently, the book focuses on wireless networks, starting with small personal area networks and progressing to very large wireless regional area networks, via local area networks dominated by WiFi technology, and finally metropolitan networks. Current personal area networks are described through Bluetooth and new types of wireless networks such as WiGig. The WiFi family continues to expand and all new members are described.

The Internet of Things is explained in a specific chapter due to its omnipresence in the literature. The forecast anticipates 100 billion connected devices by 2020. But standardized architectures and protocols are limited, which makes this field a very dense area with numerous proprietary networks. This book provides a simplified vision that ultimately makes the Internet of Things easy to understand.

Ad hoc and mesh networks are important as they have made a comeback after a long period of near hibernation. New and significant progress has been made in the field of algorithms that allows such networks to run smoothly while providing high quality service.

The last chapter discusses mobile edge computing (MEC) servers. These servers, placed close to users at the edge of the network, provide a cloud, signifying decentralization towards data centers which are much smaller than the leading cloud providers. These servers should be used to support all the associated algorithms for accessing networks, user data management, virtual machine storage and secure communication access. MEC is becoming more and more important with the massive scale of Internet traffic. Concentration of data and computing in a global cloud is becoming impractical; the world of connectivity is moving towards distributed data centers and MEC servers.

This book describes the development of wireless and mobile networks and how they will evolve in the future. The book is not exhaustive, because the field is vast and still expanding, but we hope it will be useful to the reader who wants to understand wireless networks, major innovations in the field, and current by manufacturer, operator and cloud provider actions.

Khaldoun AL AGHA
Guy PUJOLLE
Tara ALI-YAHIYA
June 2016

Acronyms

1G	first generation
2G	second generation
3G	third generation
3GPP	third-generation partnership project
4G	fourth generation
5G	fifth generation
AAA	authorization, authentication and accounting
ACK	acknowledgment
ACS	access categories
ACS	adaptive channel selection
ACS	auto configuration server
AMC	adaptive modulation and code
AMPS	advanced mobile phone system
AP	access point
ARQ	automatic retransmission request
ARU	average revenue per user
ASN	access service network
BE	best effort
BS	base stations

CA	capital expenditure
CBR	constant bit rate
CDMA	code division multiple access
CF	cyclic prefix
CI	connection identifier
CMC	connection mobility control
CN	core network
CQI	channel quality information
CQICH	channel quality indicator channel
CSG	closed subscriber group
CS	circuit switched
CSMA	carrier sense multiple access
CSN	connectivity service network
D2D	device- to-device
DCF	distributed coordination function
DCS	dynamic channel selection
DHCP	Dynamic Host Control Protocol
DL	downlink
DRA	dynamic resource allocation
DSAR	dynamic service addition request
EPC	evolved packet core
EDGE	enhanced data rates for global evolution
EPS	evolved packet system
ERT-VR	extended real-time variable rate
ETP	encapsulating tunnel payload
ETSI	European Telecommunications Standards Institute
ETSI	European Telecommunications Standards Institute
E-UTRAN	Evolved Universal Terrestrial Radio Access Network

EX-PF	exponential proportional fair
FA	foreign agents
FAP	femto access point
FBSS	fast base station switching
FCH	frame control header
FDD	frequency division duplex
FDMA	frequency division multiple access
FEC	forward error correction
FMC	fixed mobile convergence
GBR	guaranteed bit rate
GERAN	GSM/Edge Radio Access Network
GSM	global system for mobile communications
GTP	GPRS Tunneling Protocol
HA	home agent
HCCA	HCF-controlled channel access
HPU	high-priority users
HRPD	high-rate packet data
HSCSD	high-speed circuit-switched data
HSS	home subscriber server
ICIC	intercell interference coordination
IEEE	Institute of Electrical and Electronics Engineers
IETF	Internet Engineering Task Force
IKE	Internet key exchange
IMS	IP multimedia subsystem
IMT	International Mobile Telecommunications
IP	Internet Protocol
IPSec	IP Security Protocol
ISI	intersymbol interference

LAN	local area networking
LB	load balancing
LDPC	low-density parity check
LPU	low priority users
LTE	long-term evolution
LTE/SAE	long-term evolution/system architecture evolution
MAC	medium access control
MAN	Metropolitan Area Network
MEC	mobile edge computing
MBMS	Multimedia Broadcast and Multicast Service
MCS	modulation and coding scheme
MDH	macro diversity handover
MICS	Media Independent Command Service
MIES	Media Independent Events Service
MIFS	Media Independent Information Service
MIH	media-independent handover
MIMO	multiple input/multiple output
MLWDF	modified largest weighted delay first
MME	mobility management entity
MS	mobile stations
MTC	machine-type communication
MTSO	Mobile Telephone Switching Office
NAP	network access provider
NAS	non-access stratum
NGMN	next-generation mobile network
NMTS	Nordic Mobile Telephone System
nrtPS	non-real-time polling service
OAMP	operation administration maintenance and provisioning

OFDMA	orthogonal frequency-division multiple access
OPEX	operational expenditure
PAR	peak-to-average power ratio
PAN	Personal Area Network
PCEF	policy and charging enforcement function
PCI	physical cell identity
PCRF	policy and charging rules function
PDA	personal data assistants
PDU	Protocol Data Units
PHY	physical layer
PLMN	public land mobile network
PMP	point-to-multipoint
PRN	pseudo-random numerical
PS	packet scheduling
PS	packet switched
QoS	quality of service
RAN	Regional Area Network
RBC	radio bearer control
RNC	radio network controller
RRC	radio resource control
RRM	radio resource management
RSSI	received signal strength indicator
rtPS	real-time polling services
SAP	service access point
SC-FDMA	single-carrier frequency-division multiple access
SCTP	Stream Control Transmission Protocol
SF	service flow
SG	serving gateway
SIM	subscriber identity module

SINR	signal-to-interference noise ratio
SMG	special mobile group
SOHO	small office home office
SPID	subscriber profile ID for RAT/frequency priority
SS	subscriber station
TACS	total access communications system
TCP	Transmission Control Protocol
TDD	time division duplexing
TDMA	time division multiple access
TS	traffic streams
TTI	transmission time interval
TXOP	traffic opportunity
UDP	User Datagram Protocol
UE	user equipment
UGS	unsolicited grant services
UL	uplink
UMB	ultra mobile broadband
UMTS	Universal Mobile Telecommunications System
UMTS AKA	UMTS Authentication and Key Agreement
UPT	universal personal telecommunication
UTRA	universal terrestrial radio access
UTRAN	UMTS Terrestrial Radio Access Network
VLR	visitor location register
WAG	WIFI access gateway
WCDMA	Wideband Code Division Multiple Access
WiFi	wireless fidelity
WiMAX	Worldwide Interoperability for Microwave Access

1

Introduction to Mobile and Networks

The development of mobile and wireless communications was traditionally viewed as a sequence of successive generations. The first generation of analog mobile telephony was followed by the second or digital generation. The third generation enables full multimedia data transmission as well as voice communications. The fourth generation is completely Internet Protocol (IP)-based, including voice communications, and increases the throughput in parallel to these activities related to the evolution of current fourth-generation (4G) wireless technologies. There is also increased research effort on future radio access, referred to as fifth-generation (5G) radio access. Such future radio access is anticipated to take the performance and service provisioning of wireless systems a step further, providing data rates of up to 200 Mbps with wide-area coverage and up to 1 Gbps with local-area coverage. 5G technologies are being focused on as it is expected to eventually deliver approximately 10 Gbps. This can be considered as a normal evolution in response to increased user behavior, demand and quality of service (QoS) expectations.

In this chapter, we provide a brief overview of mobile and wireless networks (MWN). The objective is to present the background and context necessary for understanding subsequent chapters. We review the history of MWN, enumerate their applications and compare them in order to see the effect of such technology not only on the market drivers but also on research domain areas.

1.1. Mobile and wireless generation networks

The International Telecommunication Union (ITU) launched International Mobile Telecommunications (IMT-2000) as an initiative to cover high-speed, broadband and IP-based mobile systems featuring network-to-network interconnection, feature/service transparency, global roaming and seamless services independent of location. IMT-2000 aims to bring high-quality mobile multimedia telecommunications to a worldwide mass market by increasing the speed and ease of wireless communications, responding to problems due to increased demand to pass data via telecommunications, and providing "anytime, anywhere" services.

Two partnership organizations were born out from the ITU–IMT-2000 initiative: the Third Generation Partnership Project (www.3gpp.org) and the Third Generation Partnership Project 2 (www.3gpp2.org). The 3GPP and 3GPP2 developed their own version of 2G, 3G and later mobile systems. In parallel, the Institute of Electrical and Electronics Engineers (IEEE) was developing proper versions of the wireless networks that can be compared functionally with those of 3GPP and 3GPPP2 and their technology-based generations can be crossed with those of 3GPP and 3GPP2. Their terminologies are different but the goal is the same, which is to develop new technologies that make use of advances in the area of wireless and mobile technologies. This is why, we will summarize all the generations developed by these organizations as a path of evolution in the world of mobile and wireless networking.

1.1.1. *First generation mobile technology: 1G*

First-generation cellular networks (1G) were analog-based and limited to voice services and capabilities. Compared to today's technology, 1G technology was vastly inferior. In the late 1970s and early 1980s, various 1G cellular mobile communication systems were introduced; the first such system, the Advanced Mobile Phone System (AMPS) was introduced in the United States in the late 1970s. Other 1G systems include the Nordic Mobile Telephone System (NMTs) and the Total Access Communications System (TACS). While these systems offer reasonably good voice quality, they provide limited spectral efficiency. The evolution toward 2G was thus necessary to overcome the drawback of such technology.

1.1.2. *Second generation mobile technology: 2G*

The second-generation (2G) digital systems promised higher capacity and better voice quality than their analog counterparts. The two widely deployed 2G cellular systems are Global System for Mobile Communications (GSM) and Code Division Multiple Access (CDMA) that was originally known as American Interim Standard 95, or IS-95 and is now called cdmaOne. Both the GSM and CDMA camps formed separate 3G partnership projects (3GPP and 3GPP2, respectively) to develop IMT-2000-compliant standards based on the CDMA technology. GSM differs from 1G by using digital cellular technology, Time Division Multiple Access (TDMA) transmission methods and slow-frequency hopping for voice communication. In the United States, 2G cellular standardization process utilized direct-sequence CDMA with phase-shift keyed modulation and coding.

There was an evolution of main air interface-related enhancements to GSM: (1) higher data-rates for circuit-switched services through aggregation of several time-slots per TDMA frame with high-speed circuit-switched data (HSCSD); (2) general packet radio service (GPRS), which had efficient non-real-time packet-data traffic support. GPRS reached peak data rates of up to 140 Kbps when a user aggregated all timeslots; and (3) enhanced data rates for global evolution (EDGE) increased data rates up to 384 Kbps with high-level modulation and coding within the existing carrier bandwidth of 200 kHz.

1.1.3. *Third generation mobile technology: 3G*

Further evolution of the GSM-based systems is handled under 3GPP to define a global 3G Universal Mobile Telecommunications System (UMTS). The main component of this system is the UMTS Terrestrial Radio Access Network (UTRAN) based on Wideband Code Division Multiple Access (WCDMA) radio technology, since it uses 5 MHz bandwidth and GSM/EDGE Radio Access Network (GERAN) based on (GSM) enhanced data rates.

3GPP2 implemented CDMA2000 in the 1.25 MHz bandwidth, which increased voice and data services and supported a multitude of enhanced broadband data applications, such as broadband Internet access and multimedia downloads. This technology also doubled user capacity over cdmaOne, and with the advent of 1xRTT, packet data was available for the first time.

The 3GPP2 first introduced high-rate packet data (HRPD), termed CDMA20001xEV-DO. This standard enables high-speed, packet-switched techniques designed for high-speed data transmissions, enabling peak data rates beyond 2 Mbps. 1xEV-DO expanded the types of services and applications available to end users, enabling carriers to broadcast more media-rich content.

The 3GPP enhanced the WCDMA system, providing high-speed downlink packet access (HSDPA) that brought spectral efficiency for higher speed data services in 2001. Then, High-Speed Uplink Packet Access (HSUPA) was introduced in 2005. The combination of HSDPA and HSUPA is called HSPA. The latest HSPA is HSPA+, which resulted from adding multiple input/multiple output (MIMO) antenna capability and 16QAM (Uplink)/64QAM (Downlink) modulation. Coupled with improvements in the radio access network for continuous packet connectivity, HSPA+ allows uplink speeds of 11 Mbps and downlink speeds of 42 Mbps.

As the successor of CDMA2000, CDMA2000 1xEV-DO Release 0 provides peak speeds of up to 2.4 Mbps with an average user throughput of between 400 and 700 Kbps. The average uplink data rate is between 60 and 80 Kbps. Rel. 0 makes use of existing Internet protocols, enabling it to support IP-based connectivity and software applications. In addition, Release 0 allows users to expand their mobile experience by enjoying broadband Internet access, music and video downloads, gaming and television broadcasts.

1xEV-DO Release 0 has been revised to produce Revision A (Rev-A), which increases peak rates on reverse and forward links to support a wide-variety of symmetric, delay-sensitive, real-time, and concurrent voice and broadband data applications. It also incorporates orthogonal frequency-division multiple access (OFDMA) technology to enable multicasting (one-to-many) for multimedia content delivery. As the successor of Rev-A, 1xEV-DO Revision B (Rev-B) introduces dynamic bandwidth allocation to provide higher performance by aggregating multiple 1.25 MHz Rev-A channels.

1.1.4. *Fourth generation mobile technology: 4G*

Fourth-generation or 4G technologies allow wireless carriers to take advantage of greater download and upload speeds to increase the amount and types of content made available through mobile devices. 4G networks are using full IP solutions that deliver voice, data and multimedia content to mobile users anytime and almost anywhere. They offer greatly improved data rates over previous generations of wireless technology. Faster wireless broadband connections enable wireless carriers to support higher-level data services, including business applications, streamed audio and video, video messaging, video telephony, mobile TV and gaming.

As a step toward 4G mobile broadband wireless, 3GPP began its initial investigation of the Long-Term Evolution (LTE) standard as a viable technology in 2004. LTE offers a number of distinct advantages over other wireless technologies including increased performance attributes, such as:

– high spectral efficiency;

– very low latency;

– it supports variable bandwidths;

– simple protocol architecture;

– compatibility and interworking with earlier 3GPP releases;

– interworking with other systems, e.g. cdma2000;

– Frequency division duplex (FDD) and time division duplex (TDD) within a single radio access technology;

– efficient multicast/broadcast.

Ultra-Mobile Broadband (UMB), for the cdma2000 cellular telecommunications system, is run under the auspices of 3GPP2. The UMB cellular telecommunications system offers many new features and techniques that enable it to fulfill high expectations, and compete with other new and emerging technologies:

– data rates of over 275 Mbps in the downlink (base station to mobile) and over 75 Mbps in the uplink (mobile to base station);

– uses an OFDM / OFDMA air interface;

– uses FDD;

– possesses an IP network architecture;

– has a scalable bandwidth between 1.25 and 20 MHz (OFDM/OFDMA systems are well suited for wide and scalable bandwidths);

– supports flat, mixed and distributed network architectures.

Despite this, UMB technology was abandoned in favor of 3GPP 4G.

1.1.5. *Fifth generation mobile technology: 5G*

Studies carried out by industry players and academic into the actual use of the Internet by users led to the development of 5G. Efforts converged to fix a general view about this technology including very high data rates everywhere with low latency of end-to-end communication due to ultra-reliability and availability. Such technology is promising up to 10 Gbps and can even reach 100 Mbps in situations of indoor communication in urban or suburban areas.

The definition of the technology can be represented by the evolution of LTE Advanced itself, but some evolution in terms of utilization of spectrum, higher frequency bands and advanced multi-antenna transmission techniques, and different kinds of communications can be included in this technology such as device-to-device communication with flexible spectrum usage.

1.2. IEEE technologies

LTE is not the only solution for delivering broadband mobile services. Several proprietary solutions, particularly for fixed applications, are already on the market. There are standards-based alternative solutions that at least partially overlap with LTE, particularly for portable and mobile applications. In the near term, the most significant of these alternatives are 3G cellular systems and IEEE 802.11-based WiFi systems. In this section, we compare and contrast the various standards-based broadband wireless technologies and highlight the differentiating aspects of LTE.

1.2.1. *IEEE 802.15: WPAN*

Wireless Personal Area Networks (WPAN) or short-distance wireless networks focus on communication and interoperability among of devices operating under the WPAN. One of the first technologies based on this standard was Bluetooth, which is based on IEEE 802.15.1. However, this standard evolved to include low power consumption with a higher data rate such as Zigbee technology or standard IEEE 802.15.3 targeting a higher data rate intended for point-to-point close-proximity communication including kiosk downloading and intradevice communication just like wireless data centers and wireless backhauling. Accordingly, some forums were founded within the IEEE working group itself in order to provide a higher speed ultra-wideband (UWB) for applications, which involved imaging and multimedia. As a result of all these standards and technologies based on these standards, WiMedia Alliance was created to be responsible for the adoption, regulation, standardization and multi-vendor interoperability of UWB technologies.

1.2.2. *IEEE 802.11: WLAN*

The Wireless Fidelity (WiFi)-based-system is used to provide broadband wireless. It is based on the IEEE 802.11 family of standards and is primarily a local area networking (LAN) technology designed to provide in-building broadband coverage. Current WiFi systems based on IEEE 802.11a/g support a peak physical-layer data rate of 54 Mbps and typically provide indoor and outdoor coverage over a few 1000 m^2, making them suitable for enterprise networks and public hot spot scenarios such as airports and hotels.

WiFi offers remarkably higher peak data rates than 3G systems, primarily since it operates over a larger 20 MHz bandwidth. The inefficient Carrier Sense Multiple Access (CSMA) protocol used by WiFi, along with the interference constraints of operating in the license-exempt band, is likely to significantly reduce the capacity of outdoor WiFi systems. Further, WiFi systems are not designed to support high-speed mobility.

A major benefit of WiFi over World Wide Interoperability for Microwave Access (WiMAX) and 3G is the wide availability of terminal devices. A vast majority of laptops have a built-in WiFi interface. WiFi interfaces are now also being built into a variety of devices, including Personal Data Assistants (PDAs), cordless phones, cellular phones, cameras and media players. This will enable an easy use of broadband network services using WiFi. As with 3G, the capabilities of WiFi are being enhanced to support even higher data rates and to provide better QoS support. In particular, using multiple-antenna spatial multiplexing technology, the IEEE 802.11n standard supports a peak layer-2 throughput of at least 100 Mbps. It is expected that MIMO antennas will use multiple antennas to coherently resolve more information than possible using a single antenna.

1.2.3. *IEEE 802.16: WMAN*

WiMAX IEEE 802.16 standard for the global deployment of broadband Wireless Metropolitan Area Networks is available in two versions: fixed and mobile. Fixed WiMAX, which is based on IEEE 802.16-2004, is ideally suited for delivering wireless, last-mile access for fixed broadband services. It is similar to digital subscriber line or cable modem services. Mobile WiMAX, which is based on the IEEE 802.16e standard, supports both fixed and mobile applications while offering users improved performance, capacity and mobility.

Mobile WiMAX provides higher data rates with OFDMA support and introduces several key features necessary for delivering mobility at vehicular speeds with QoS comparable to broadband access alternatives. Several features that are used to enhance data throughput are: Adaptive Modulation and Coding (AMC), Hybrid Automatic Repeated Request (HARQ), fast scheduling and bandwidth efficient handover. Mobile WiMAX is currently TDD operating at 2.5 GHz. Mobile WiMAX has higher tolerance to multipath and self-interference and provides orthogonal uplink multiple access with frequency selective scheduling and fractional frequency reuse.

1.2.4. *IEEE 802.21: MIHS*

Media Independent Handover Service (MIHS) (standard IEEE 802.21) specified interhandover management through different technologies regardless of type or family of standardization (3GPP or IEEE). It provides a cross-layer solution that triggers the handover at the most appropriate time when moving from one network to another. The aim is to ensure continuity of service when a user is changing its home network to another network which is using different technology. The ultimate goal is to ensure the QoS of an ongoing communication anytime and anywhere. The standard has defined many services and primitives that allow communication that propagates multilayer information from a physical layer to a higher layer so the higher layer is aware of what is going on in the lower layer, is thus handover occurs at the best time.

1.2.5. *IEEE 802.22: WRAN*

This group is mainly working on "enabling broadband wireless access using cognitive radio technology and spectrum sharing in white spaces". Its work focuses mainly on regional networks and this is why the standard is called Wireless Regional Area Networks as they use a cognitive method to detect the white space in TV broadcast bands and reuse them again without causing any interference to the licensed users in the same band. The standard specifies operation in bands that allow spectrum sharing where the communication devices may opportunistically operate in the spectrum of the primary service such as 1,300–1,750 MHz, 2,700–3,700 MHz and the VHF/UHF TV broadcast bands between 54 and 862 MHz.

1.3. Conclusion

The move from analog to digital has paved the way for new generations of WAN. All technologies, regardless of standardization family, IEEE or 3GPP or 3GPP2, are trying to fully IP-orient networks, and this is the reason behind the fast evolution that we are witnessing.

Progress and the fast evolution of wireless networks is due to the evolution of users' demands and expectations, as well as the evolution of the nature of traffic that the Internet is experiencing through different kinds of applications. Such traffic is very greedy in terms of data rate and has strict requirements in terms of QoS parameters. Such progress could not occur if there were no advanced modulation technologies in the physical layer making it possible to have such diversity in current wireless technologies.

Finally, industrial and academic research communities, even though they have had different concepts and points of view of how technology can be implemented and deployed, have been a principal factor in network development.

1.4. Bibliography

1.4.1. *Standards*

3GPP2 TSG C.S0084-001-0 v2.0, Physical Layer for Ultra Mobile Broadband (UMB) Air Interface Specification.

IEEE, STANDARD 802.16E-2005. PART 16 – Air interface for fixed and mobile broadband wireless access systems–Amendment for physical and medium access control layers for combined fixed and mobile operation in licensed band, December 2005.

1.4.2. *Selected bibliography*

[AL 05] AL AGHA K., PUJOLLE G., VIVIER G., *Réseaux de mobiles et réseaux sans fil*, 2nd ed., Eyrolles, 2005.

[AND 03] ANDERSON H.R., *Fixed Broadband Wireless System Design*, Wiley, 2003.

[AND 07] ANDREWS J.-G., GHOSH A., MUHAMMED R., *Fundamentals of WiMAX*, Prentice Hall, 2007.

[CLA 00] CLARK M.P., *Wireless Access Networks: Fixed Wireless Access and WLL Networks-Design and Operation,* Wiley, 2000.

[ELB 02] ELBERT B.R., *The Satellite Communication Ground Segment and Earth Station Handbook*, Artech House, 2002.

[ERI 15] ERICSSON, 5G Radio Acess, White Paper, Uen 284-23-3204 Rev B, February 2015.

[GAG 03] GAGNAIRE M., *Broadband Local Loops for High-Speed Internet Access,* Artech House, 2003.

[GI 08] GI LEE B., CHOI S., *Broadband Wireless Access & Local Networks: Mobile Wimax and Wifi,* 1st ed., Artech House Publishers, 2008.

[GIA 07] GIAMBENE G., *Resource Management in Satellite Networks: Optimization and Cross-Layer Design*, 1st ed., Springer, 2007.

[GOL 05] GOLDSMITH A., *Wireless Communications*, Cambridge University Press, Cambridge, 2005.

[HAL 02] HALONEN T., ROMERO J., MELERO J., *GSM, GPRS and EDGE Performance: Evolution Toward 3G/UMTS*, John Wiley & Sons, 2002.

[HAL 06] HALOMA H., TOSKALA A., *HSDPA/HSUPA for UMTS: High Speed Radio Access for Mobile Communications,* Wiley, 2006.

[IBE 02] IBE O.C., *Fixed Broadband Wireless Access Networks and Services*, Wiley, 2002.

[JOH 08] JOHNSON C., *Radio Access Networks for UMTS: Principles and Practice*, Wiley, 2008.

[KAA 05] KAARANEN H., AHTIAINEN A., LAITINEN L. *et al.*, *UMTS Networks: Architecture, Mobility and Services*, Wiley, 2005.

[MAR 82] MARAL G., BOUSQUET M., PARES J., *Les systèmes de télécommunications par satellite*, Masson, 1982.

[MIN 02] MINOLI D., *Hotspot Networks: WiFi for Public Access Locations*, McGraw-Hill, 2002.

[PAN 03] PANDYA R., *Introduction to WLLs*, Wiley, 2003.

[PAR 07] PARK Y., ADACHI F., *Enhanced Radio Access Technologies for Next Generation Mobile Communication*, 1st ed., Springer, 2007.

[PRA 02] PRATT T., BOSTIAN C.W., ALLNUTT J.E., *Satellite Communications*, Wiley, 2002.

[SMI 00] SMITH C., *LMDS: Local Multipoint Distribution Service,* McGraw-Hill, 2000.

[STE 92] STEELE R., *Mobile Radio Communications*, Pentech Press, 1992.

[WEE 02] WEBB W., *Introduction to Wireless Local Loop*, Artech House, 2002.

1.4.3. *Websites*

http://www.3gpp.org/technologies/keywords-acronyms/98-lte

http://www.ieee802.org/15/pub/TG3c.html

www.ieee802.org/11/

www.ieee802.org/21/

www.ieee802.org/22/

www.wi-fi.org/

www.wimedia.org/

2

Mobile Networks

The standardization of GSM-based systems has its roots in the 1980s, when the "Group Special Mobile" was created within the *Conference Européenne des Postes et Télécommunications*, whose task was to develop a unique digital radio communication system for Europe, at 900 MHz. The system has experienced extensive modifications to fulfill the increasing operator and cellular user demands. The main GSM system development between 1990 and 2000 was conducted by the European Telecommunications Standards Institute, Special Mobile Group and its technical subcommittees, as well T1P1, which was responsible for the PCS1900 MHz specifications in the United States.

Further evolution of the GSM-based systems was handled under the 3GPP, a joint effort of several standardization organizations around the world to define a global 3G Universal Mobile Telecommunication System (UMTS) cellular system. The main components of this system are UTRAN, based on wideband code division multiple access (WCDMA) radio technology, and GSM/EDGE radio access network (GERAN), based on GSM/EDGE radio technology.

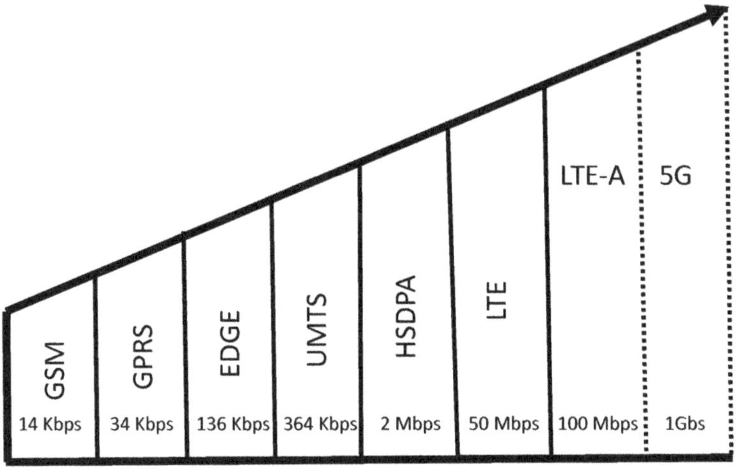

Figure 2.1. *2G and 3G technologies*

2.1. Cellular network

Mobile radio communication introduces some challenges related to radio resources. There are several problems related to the following elements:

– *signal strength*: the signal strength between the base station and the cell phone must be sufficiently high to maintain the communication. There are several factors that may influence the signal level (distance from the base station, interfering signals, etc.);

– *fading*: the different effects of signal propagation can cause disruptions and errors. It is important to consider these factors during the construction of a cellular network. To ensure quality communication and to avoid interference, cellular networks use signal power control techniques.

We want the power of the received signal to be sufficiently greater than the background noise. For example, when the cell phone moves away from the base station, the received signal is attenuated. Conversely, the effects resulting from the reflection, diffraction and dispersion can change the signal strength even if the cell phone is close to the base station. It is also important to reduce the power of the signal transmission from the cell to avoid interference with neighboring cells but also to protect health and save energy.

2.1.1. *Radio interface*

As radio resource is scarce, data multiplexing methods have been used to optimize its use. Frequency Division Multiple Access (FDMA) is the multiple access method most frequently used. This technique is the oldest, and it differentiates users with a simple frequency differentiation. Each user has an associated predetermined frequency. The implementation of this technology is quite simple.

Time Division Multiple Access (TDMA) is based on the distribution of resources over time. Each frequency is divided into time intervals. Each user sends or transmits in a range of concrete time whose periodicity is defined by the duration of the frame. In this case, to listen to N users, the receiver has to consider the time interval T associated with this user. Unlike FDMA, multiple users can transmit on the same frequency.

Code Division Multiple Access (CDMA) is the dedicated 3G access method. It is based on code division. It spreads the spectrum by allocating a code to each communication. Each user is differentiated from other users with a code that was allocated early in his communication and which is orthogonal to other users' related codes. In this case, to listen to user N, the receiver need only multiply the signal received by code N for the user.

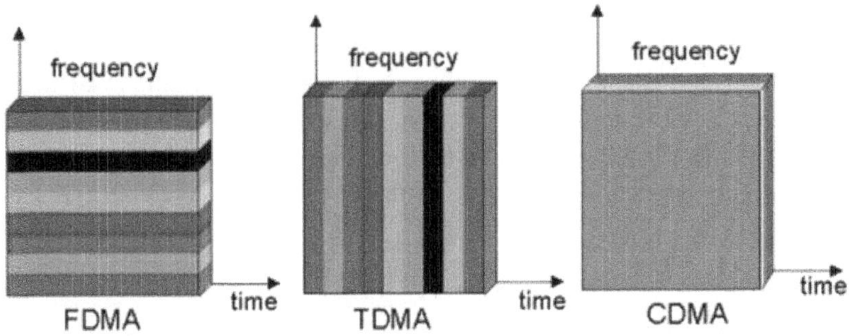

Figure 2.2. *The different access methods*

However, the traffic in the downlink and uplink can be merged by time division duplex (TDD) or frequency division duplex (FDD) multiplexing.

Orthogonal Frequency Division or OFDM is another method of modulation that is used in the most recent technology just like WLAN based on IEEE 802.11a, IEEE 802.11g, UWB, WiMAX, LTE and 4G, etc. It provides high spectral efficiency, resilience to radio frequency interference and lower multi-path distortion. It is important first to understand the frequency division multiplexing (FDM) that extends the principle of single carrier modulation through the use of multiple subcarriers in the same channel. The data rate divided between the various subcarriers will be sent in one channel but different subcarriers. If the set of the subcarriers is orthogonal to each other, we add the word orthogonal to the technology. The guardbands necessary to allow individual demodulation of subcarriers in a FDM system would no longer be necessary. The use of orthogonal subcarriers allows subcarriers' spectra to overlap, improving spectral efficiency. As long as orthogonality is maintained, it is still possible to recover the individual subcarriers' signals despite their overlapping spectrums.

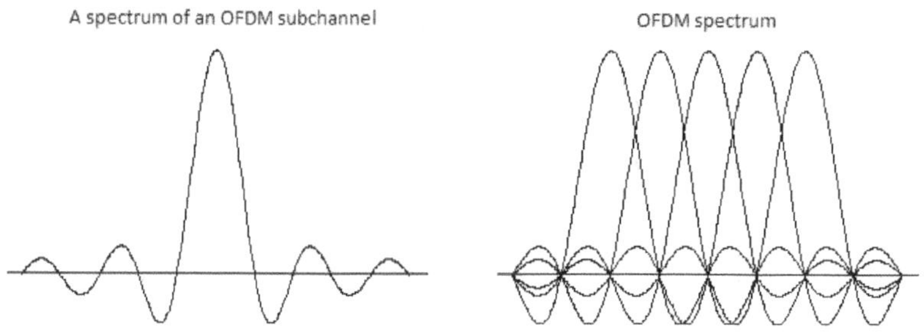

Figure 2.3. *An OFDM spectrum for a single subchannel (left), and five carriers (right)*

2.1.2. *Cell design*

A cellular network is based on the use of low-power transmitters (≈ 100 W). The range of such a transmitter is limited to a geographical area divided into smaller areas called cells. Each cell has its own transceiver (antenna) under the control of a base station and each cell has a certain frequency range. To avoid interference, adjacent cells should not use the same frequencies, however two cells far enough away from each other can use the same frequencies. The cells are designed in hexagonal form to facilitate the decision to change a cell range. If the distance between all cell transmitters is the same, then it is easy to standardize the time when a mobile node should change cell. In practice, the cells are not quite hexagonal because of the different topology and the propagation conditions, etc.

Another important choice in the construction of a cellular network is the minimum distance between two cells operating in the same frequency band to avoid interference. For this, the cell organization could follow different patterns. If the pattern contains N cells, each could use K/N frequencies where K is the number of frequencies allocated to the system. The advantage of frequency reuse is to increase the number of users in the system using the same frequency band. Where the whole frequency system is at maximum capacity, i.e. all frequencies are used, there are techniques that support new system users. The addition of new channels, frequency borrowing from neighboring cells and cell division are techniques that increase system capacity. The general principle is to have micro, picocells in high-density areas to allow a greater frequency reuse in a geographic area with a high population.

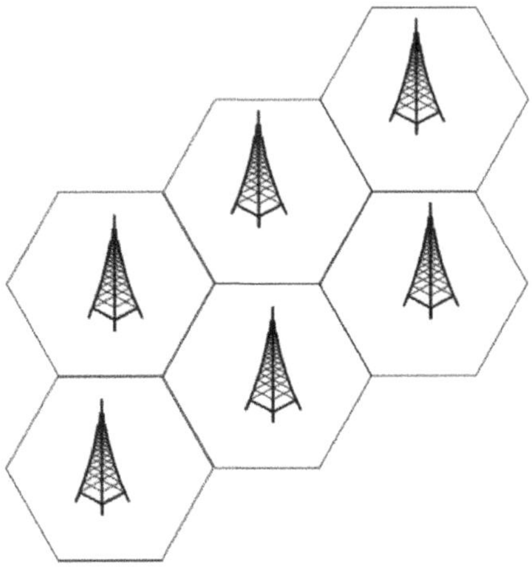

Figure 2.4. *Cell design*

2.1.3. *Traffic engineering*

Traffic engineering has been developed for the circuit switching based network design. In the context of cellular networks, it is essential to know how to plan the network to accept the maximum number of calls. One of the basic cellular network designs defines the degree of call blocking and also how to manage it. In other words, if a call is blocked, it will be put on hold and it will be necessary to define what the average waiting time is. Knowing the starting capacity of the system (number of channels), we can determine the probability of blocking and average waiting time for a blocked call. What complicates this traffic engineering in cellular networks is user mobility. In a cell, in addition to the novel calls there will be calls transferred by neighboring cells. Another parameter which further complicates the model is that the system should also combine or accept both phone calls and data traffic calls.

2.2. Principles of cellular network functionalities

A cellular network is generally composed of:

– *a Base station (BS)*: which is the center of the cell; a BS includes an antenna, a controller and a number of transceivers. It allows communications over channels assigned to the cell. The controller is used to manage the appeal process between a mobile and the rest of the network. The BS is connected to Mobile Telephone Switching Office (MTSO). Two types of channels are established between the cell phone and the BS: control channels and traffic channels. Control link the cell phone with the BS and exchange information necessary to establish and maintain connections. The traffic channels are used to transport traffic (voice, data, etc.).

– *an MTSO*: manages several BSs generally associated with a wired network. It is responsible for establishing connections between mobile nodes. It is also connected to the wired telephone network and is thus able to establish connections between fixed and mobile nodes. The MTSO is responsible for the allocation of channels to each call; it is also responsible for handovers and billing information. The call process involves the following functions:

- *initializing a mobile node*: when a mobile node is turned on, it will scan the existing frequency and select the strongest channel with this frequency. Each cell regularly broadcasts control information on the band frequency corresponding to the control channel. Then the cell phone centers the identification phase after handshakes with the BS. This phase occurs between the cell phone and the MTSO. The cell phone is identified following authentication and its geographical position is registered. It continues to regularly scan the frequency spectrum and decides to change base station if it has a stronger signal than the previous cell;

- *call initiated by a phone*: the mobile node verifies that the call channel is free according to information sent by the base station on the control channel on the downlink. The cell phone can then send the call number on the control channel to the base station so it can transmit its request to the MTSO;

- *call notification*: according to the call number received, the switching center attempts to connect to the base stations concerned with the number called and sends a call notification (paging) to the cell phone. Notification of the call is transmitted by the base stations in the call control channel in the downlink;

- *call acceptance*: once the cell phone knows its number in the channel call control, it responds to the base station which will relay the message to the switch center that will establish a circuit between the base stations and the caller by a third party. The switch will also select one traffic channel available for each of the relevant cells and communicate information on this call with the base stations. The cell phone will then synchronize on traffic channels selected by the base stations;

- *active communication*: is the voice traffic exchange process or data exchanged between the cell phone calling and the one called. This is ensured by the two base stations and the switching center;

- *call blocking*: if all the traffic from a base station channels is occupied, the cell phone will try to repeat the call a number of times. In case of failure, a busy tone signal is returned to the user;

- *call termination*: at the end of a call, the switching center informs the base station to release the channel. This action is also important for billing;

- *abandon call*: during a call, if the base station fails to maintain a good level of signal (interference, low signal strength, etc.), it changes the traffic channel of the cell phone and notifies the switching center.

- *call between landline and mobile*: the mobile switching center being connected to the fixed network, it is then able to establish communications between these two networks. It can also attach another mobile switching center through the fixed network.

- *handoff (handover)*: when the cell phone discovers a control channel in which the signal strength is stronger than the current cell, the network will automatically transfer the call to another cell in a transparent way. The decision is made according to the measured power of the signal of the BS with this mobile node. In general, the station calculates an average of a certain window of time to eliminate rapid fluctuations resulting from multipath effects. Various techniques can be used to determine the mobile transfer time. This transfer can be controlled either by the network or by the cell phone. The simplest handover decision technique is the one that initiates the transfer when the cell phone detects a stronger signal than that of the cell where it is located.

2.3. 1G networks

1G cellular networks, such as CT0/1 (cordless telephone) for wireless networks and Advanced Mobile Phone Service (AMPS) for mobile networks is characterized by analog communications. Few first network cell phones still exist. AMPS was the first most widespread system in the United States around 1980. It was also deployed in South America, Australia and China. In Northern Europe, the Nordic Mobile Telecommunication System network has been developed. In the United Kingdom, the Total Access Communication System and France Radiocom 2000 were deployed. All these 1G cellular networks were analog and used frequency bands between 450 and 900 MHz.

2.4. 2G networks

The 2G cellular networks such as digital enhanced cordless telecommunication (DECT) for wireless and GSM for cell phones are characterized by digital communications unlike 1G networks. During the 1990s, several digital technologies have been developed:

– GSM developed in Europe runs at 900 MHz;

– the dynamic channel selection (DCS) 1,800 equivalent to GSM but operating at higher frequencies (1,800 MHz);

– 1900 personal communication system (PCS) and D-AMPS (digital AMPS) have been developed in the United States;

– the Pacific or Personal Digital Cellular (PDC) system developed in Japan.

GSM, D-AMPS is also called IS-136, based on the TDMA access method, while the PCS 1900 also called cdma One or IS-95 and relies on CDMA technology.

In this generation, simple data transmission is possible in addition to voice, but the data rate remains low with less than 10 kbps and it is impossible to deploy multimedia services. High-speed circuit switched data (HSCSD) and General Packet Radio Services (GPRS) have allowed the increase in the data rate in 2G networks. These technologies are also known as 2.5 generation cellular networks. GPRS unlike HSCDC uses packet switching, thus optimizing the use of the radio resource for data transmission of sporadic nature. The theoretical data rate is 120 kbps, while the real data rate does not exceed 30 kbps. This generation could not meet the needs of cell phone users who wanted to use multimedia services comparable to fixed networks. The evolution of the GPRS network resulted in the EDGE or Enhanced GPRS (EGPRS), which improved the reliability and throughput of data communications. This network is usually known as 2.75G or 3G, depending on implementation. This is a simple evolution of GSM/GPRS to obtain data rate of 130 and 60 kbps in transmission.

Mobility management is generally carried out using two data bases: the Home Location Register (HLR), which updates the subscriber's data, and the Visitor Location Register (VLR), which manages the customer in the cell visited. The network due to these two components can provide mobile node location management to route calls and also provide handover or handoff. These cellular networks allow high mobility of the terminal but low personal mobility, resulting in the possibility of using the Subscriber Identity Module (SIM) card in any terminal. Recall that personal mobility is the ability to change terminals while retaining a working environment. We find this mobility in Universal Personal Telecommunication (UPT) networks.

2.5. 3G networks

3G cellular networks operate around the 2 GHz frequency band. They provide a range of multimedia services fixed and mobile users with a QoS comparable to that for fixed networks. The International Telecommunication Union has selected five standards for 3G mobile: (Wideband CDMA) W-CDMA, TD-CDMA and TD-SCDMA used in the European Universal Mobile Telecommunication System, CDMA2000, EDGE and 3G DECT. IMT-2000 networks are intended to include a global roaming, a range of broadband services such as video and the use of a single terminal in different wireless networks.

Another goal is for fixed and mobile services to provide consistent transparency to the user. These networks provide a global mobility that includes mobility terminal, personal mobility and service mobility. The concept of the Virtual Home Environment (HEV) is then developed to support this mobility service. In addition to the higher data rates, the overall mobility is another major difference to 2G networks.

UMTS based on the W-CDMA access method theoretically allows a data rate of up to 1.920 Mbps, but in late 2004 the data rate offered by operators rarely exceeded 384 kbps. This rate is well above the GSM base data rate (9.6 kbps). UMTS-TDD is based on the TD-CDMA access method, which is not compatible with UMTS. The 3G network developed in China is based on time division-synchronous code division multiple access (TD-SCDMA), a local standard, to avoid paying for the copyrights of other 3G standards.

CDMA2000 1x, CDMA2000 1xEV-DO and CDMA2000 1xEV-DV are direct successors of 2G CDMA (cdmaOne IS-95), and these are 3GPP1 standards. CDMA2000 1x, known as 1X, 1xRTT, IS-2000, CDMA2000 1X, 1X, or cdma2000, doubles the capacity of the votes compared to IS-95. Data rate transmission can reach 144 kbps. 1xRTT is considered as 2.5G or 2.75G or 3G depending on the implementation. CDMA2000 3x was specified on another frequency band. This standard has not been deployed. Finally, 1xEV-DO 4 IS-856 or 1xEV-DV5 is used to increase the data transmission rate and support mobile video.

In the High Speed Protocol Access (HSPA) family, which should evolve UMTS network toward a high data transmission rate, there are the HSDPA, HSUPA and HSOPA protocols which are the successors of UMTS. High-Speed Uplink Packet Access (HSUPA) can handle a rate of 5.76 Mbps. High-speed downlink protocol access (HSDPA) in the first phase of its development supported up to 14 Mbps. In the second phase of it could support up to 28.8 Mbps through MIMO technology and beamforming. High Speed OFDM Packet Access (HSOPA), the successor of HSDPA, reached 100 Mbps downlink and 50 Mbps in the uplink through the OFDMA access technology. It competes directly with IEEE technologies such as WiMAX. HSOPA is a new air interface that is incompatible with W-CDMA and therefore with developments in previous 3G networks.

The final evolution of 3G consists of Long-Term Evolution (LTE), which is promoted by the 3GPP to describe commercial technology corresponding to HSOPA, that is to say, the arrival of the mobile radio technologies using OFDMA. The ultra mobile broadband (UMB) promoted by the 3GPP2 aims to succeed cdma2000, but this solution is rapidly losing ground compared to LTE.

LTE began to be marketed in 2010 and most countries have deployed this technology to handle the enormous data rates from smartphones and tablets. Feedback is that the peak rates are rarely achieved. One can say that the conventional flow rates are in the order of those of ADSL, that is to say, several megabits per second. Mobile operator advertisements call this 4G standard, but in reality it belongs to 3G. LTE should be called 3.9G. We are not far from 4G, and only the voice is still non-VoIP. It is possible to transmit VoIP over LTE, but through the data channel: this is the Voice over LTE (VoLTE), as we will discuss Chapter 3.

Recall that the LTE network includes the radio part, or E-UTRAN, and the network part, with the evolved packet core (EPC). In E-UTRAN, the control functions were built into the eNodeB when they were pushed into the Radio Network Controller in UMTS networks. This support is changed in the 5G, where control schemes mainly perform in the Cloud and usually in a local data center. The EPC is definitely IP-based.

LTE network performances have improved since the launch of these networks in 2009. The downstream speeds can theoretically reach 326.4 Mbps using a MIMO with four antennas. The flow pic rate is 86.4 Mbps. The roundtrip time, is much better than that of UMTS and can drop to 10 ms in a good case.

2.6. 4G networks

LTE Advanced and WiMAX Phase 2 mark the real start of 4G. LTE Advanced follows LTE, with IP compatibility. WiMAX Phase 2 also belongs to the 4G family and is IP-compatible in nature. Many functions have been revised compared to WiMAX Phase 1. 4G is therefore fully compatible with the IP world. However, WiMAX Phase 2 is not a success story and should disappear leaving LTE Advanced (LTEA) as the only 4G solution.

Many new items are included in the 4G represented by the LTE Advanced technology. The highlights are:

– the relay nodes that achieve an Enhanced NodeB (eNodeB) antenna or eNB via a pick-up point. This solution was proposed primarily in the context of WiMAX Phase 2 but was taken over by LTE Advanced;

– MIMO solutions are used in the transmission to the terminal and in the opposite direction. MIMO is described in Chapter 9;

– bandwidths used range from 20 to 100 MHz, which allows increased rates up to a factor of five when necessary;

– the air interface is optimized based on the location of the terminal and interferences. In other words, as in the 3G+ and LTE technologies, a set of tools speed up the rates as soon as the terminal is located in an area where a good quality signal is received and throughput degrades when the client is in a poor environment or when the terminal is highly mobile;

– 4G selected a compatibility with WiFi environments to take advantage of these networks for important flows. The concept of the femtocell is much more developed in this generation;

– cognitive radio, which we introduce later on, is increasingly used. This allows for both greater bandwidth and better use of the spectrum;

– the management and control schemes of the network are much more automated and become autonomic, i.e. autonomous and spontaneous. A major effort is made to obtain an automatic piloting system for configuration, security, availability and monitoring;

– the encoding of the information is improved for even higher compression rates than 3G+. Error corrections are improved, but are rather similar to techniques used in 3G+;

– interference management is greatly improved, so interference is virtually eliminated. These technologies use directive antennas (beamforming), which by their directivity can determine the location of the end machine;

– the techniques used on the air interface may be asymmetric, for example a technical OFDMA on the downlink part and SC-FDMA on the uplink part.

Returning to the purely technical part, LTE-A offers significantly higher data rates through the aggregation of different carriers (Carrier Aggregation) and the disappearance of the GSM digital word to become completely VoIP. The carrier aggregation brings together carriers that are not contiguous. With a spectrum width of 100 MHz, it is possible to obtain a flow rate of about 1 Gbps. MIMO technology can also be extended with eight antennas. Radio relays, that we describe in Chapter 8, will also be much more widespread. The “small cell”, which we will examine in Chapter 7, will be multiplied to allow a multiplication of the capacity.

2.7. 5G networks

5G is not yet defined, but many texts are already available. The basic technology is no different from that of 4G. The differences are more architectural: the introduction of Cloud to control mobile network equipment, the strong use of virtualization and the ability to connect billions of “things”.

Let us start with the introduction of Cloud solutions to support network control schemes. The Cloud provides computing power and information retrieval that can make decisions taking into account a large number of parameters. For example, an algorithm can decide about a handover rather than considering only the signal strength and a few local parameters. For example, when the signal reception on a new antenna is higher than on the current antenna, it may be best to take into account parameters such as load of the cells, evaluation of the load of the transit networks, the energy savings, the type of application in transit, etc. Therefore, it is possible to greatly refine the decision. The downside is of course the time taken to go back and forth between the network nodes that implement the decisions and the place where decisions are made. This approach is similar in all respects to that of Software-Defined Networks (SDNs). Here, since real time is often necessary, it could be important for the data center in which the algorithms are executed be close to the network equipment. One can speak of distributed Cloud or Local Cloud. This solution will be described in Chapter 13.

The second main direction is virtualization. We detail network virtualization techniques in Chapter 13. The main idea of virtualization is for network equipment to be shared between several operators and between users. For example, a VLR or a 3G/4G antenna or a WiFi hotspot can be shared between multiple users. In this solution, each user has the impression that the network belongs to him. Indeed, the user may modify control and management software. The advantage is much better use and sharing of equipment between several operators. Virtualization may allow coexistence of heterogeneous networks, since the networks are isolated from each other.

The most visible part of 5G comes from the connection of billions of things. This is called Internet of Things. The forecast is 100 billion things by 2022. We will discuss this new environment in Chapter 5.

2.8. Bibliography

[AGG 04] AGGELOU G., *Mobile Ad Hoc Networks: From Wireless LANs to 4G Networks*, 1st ed., McGraw-Hill Professional, 2004.

[BEK 01] BEKKERS R., *Mobile Telecommunications Standards: UMTS, GSM, TETRA*, Book News, 2001.

[BLA 98] BLACK U.D., *Second Generation Mobile and Wireless Networks*, Prentice Hall, 1998.

[BRA 02] BRAND A., AGHVAMI H., *Multiple Access Protocols for Mobile Communications: GPRS, UMTS and Beyond*, John Wiley & Sons Ltd, 2002.

[BRA 04] BRAITHWAITE C., SCOTT M., *UMTS Network Planning and Development: Design and Implementation of the 3G CDMA Infrastructure*, Newnes, 2004.

[BUR 02] BURKHART J., *Pervasive Computing: Technology and Architecture of Mobile Internet Applications*, Addison Wesley, 2002.

[CAS 01] CASTRO J.P., *The UMTS Network and Radio Access Technology: Air Interface Techniques for Future Mobile Systems*, John Wiley & Sons Ltd, 2001.

[CAS 02] CASTRO J.P., *The UMTS Network and Radio Access Technology*, Wiley, 2002.

[CAS 04] CASTRO J.P., *All IP in 3G CDMA Networks: The UMTS Infrastructure and Service Platforms for Future Mobile Systems*, Wiley, 2004.

[COR 07] CORREIA L.M., *Mobile Broadband Multimedia Networks: Techniques, Models and Tools for 4G*, 1st ed., Academic Press, 2007.

[ESM 02] ESMAILZADEH R., Nakagawa M., *TDD-CDMA for Wireless Communications*, Artech House, 2002.

[FAG 01] FAGGION N., *Le GPRS : du WAP à l'UMTS*, Dunod, 2001.

[GAR 97] GAR V.K., SMOLIK K., WILKES J.E., *Applications of CDMA in Wireless/Personal Communications,* Prentice Hall, 1997.

[GAR 99] GARG V., *IS-95 CDMA and cdma2000*, Prentice Hall, 1999.

[HAL 02] HALONEN T., *GSM, GPRS and EDGE Performance – Evolution Towards 3G/UMTS*, Wiley, 2002.

[HAL 03] HALONEN T., ROMERO J., MELERO J., *GSM, GPRS and EDGE Performanc*e, 2nd ed., John Wiley & Sons, Ltd. 2003.

[HAN 02] HANZO L., WONG C.H., YEE M.S., *Adaptive Wireless Transceivers: Turbo-Coded, Turbo-Equalized and Space-Time Coded TDMA, CDMA, and OFDM Systems*, Wiley, 2002.

[HEI 99] HEINE G., *GSM Networks: Protocols, Terminology and Implementation*, Artech House, 1999.

[HOL 01] HOLMA H., TOSKALA A., *UMTS : Les réseaux mobiles de troisième génération*, Osman Eyrolles Multimédia, 2001.

[HOL 04] HOLMA H., TOSKALA A., *WCDMA for UMTS*, 3rd ed., John Wiley & Sons, Ltd, 2004.

[KAR 02] KARIM M.R., SARRAF M., *W-CDMA and cdma2000 for 3G Mobile Networks*, McGraw-Hill, 2002.

[LEE 01] LEE C.-C., GOULD P., *GSM, cdmaOne and 3G Systems*, Raymond Steele, John Wiley & Sons Ltd, 2001.

[LEE 02] LEE S., *Spread Spectrum CDMA: IS-95 and IS-2000 for RF Communications*, McGraw-Hill, 2002.

[LIN 00] LIN Y.B., CHLAMTAC I., *Wireless and Mobile Network Architectures*, Wiley, 2000.

[MAC 97] MACARIO R.C.V., *Cellular Radio, Principles and Design*, 2nd ed., Macmillan, 1997.

[MAN 02] MANDYAM G.D., LAI J., *Third Generation CDMA Systems for Enhanced Data Services*, Academic Press, 2002.

[MOR 06] MORISIO M., TORCHIANO M., *Developing Services for the Wireless Internet*, 1st ed., Springer, 2006.

[MUL 95] MULLER N.J., TYKE L.L., *Wireless Data Networking*, Artech House, 1995.

[MUR 00] MURATORE F., *UMTS: Mobile Communications for the Future*, Wiley, 2000.

[OJA 98] OJANPERA T., PRASAD R., *Wideband CDMA for Third Generation Mobile Communications*, Artech House, 1998.

[PRA 98] PRASAD R., *Universal Wireless Personal Communications*, Artech House, 1998.

[ROS 03] ROSENBERG R., KEMP S., *CDMA Capacity and Quality Optimization*, McGraw-Hill, 2003.

[SHA 48] SHANNON C., "A mathematical theory of communication", *Bell System Technical Journal*, vol. 27, pp. 379–423, 623–656, 1948.

[SHO 06] SHOREY R., ANANDA A., CHAN M.C. *et al.*, *Mobile, Wireless, and Sensor Networks: Technology, Applications, and Future Directions*, 1st ed., Wiley–IEEE Press, 2006.

[SIN 01] SINGHAL S., BRIDGMAN T., SUYRANARAYANA L. *et al.*, *The Wireless Application Protocol: Writing Applications for the Mobile Internet*, Addison Wesley, 2001.

[SMI 06] SMITH C., *3G Wireless Networks*, 2nd ed., McGraw-Hill Osborne Media, 2006.

[STE 01] STEELE R., *GSM, cdmaOne and 3G Systems*, Wiley, 2001.

[TIS 00] TISAL J., *The GSM Network: The GPRS Evolution: One Step Towards UMTS*, Wiley, 2000.

[TOH 01] TOH C.K., *Ad Hoc Mobile Wireless Networks: Protocols and Systems*, Prentice Hall, 2001.

[TSE 05] TSE, VISWANATH P., *Fundamentals of Wireless Communication*, Cambridge University Press, 2005.

[VAN 04] VANGHI V., DAMNJANOVIC A., VOJCIC B., *The cdma2000 System for Mobile Communications*, Prentice Hall, 2004.

[VIT 95] VITERBI A.J., *CDMA Principles of Spread-Spectrum Communications*, Addison Wesley, 1995.

3

Long-Term Evolution

The development of Long-Term Evolution (LTE) is now finished. LTE focuses on enhancing the Universal Terrestrial Radio Access (UTRA). LTE mobile broadband is popularly called a 4G developed by the Third Generation Partnership Project (3GPP) and adopted by the European Telecommunications Standards Institute. In fact, LTE is not 4G but a member of 3G and should be called 3.9G but it is easier to talk about 4G. The main difference concerns the voice: 4G is a full-IP technology and the voice is transferred through a VoIP technology. In the LTE, mobile operators are using GSM for voice communications. Obviously, it is possible to get VoIP within 3.9G technology using the data band: this is called VoLTE.

LTE provides an average user throughput of three to four times the Release 6 High Speed Download Packet Access (HSDPA) levels in the downlink (100 Mbps) and two to three times the HSUPA levels in the uplink (50 Mbps). In 2007, 3G Evolved Universal Terrestrial Radio Access progressed from the feasibility study stage to the first approved technical specifications. By the end of 2008, the specifications were sufficiently stable for the first wave of LTE equipment in Release 8. However, some small enhancements were introduced in Release 9, that was functionally frozen in December 2009. The motivation for 3GPP Release 8 was;

– the need to ensure the continuity of competitiveness of the 3G system for the future;

– user demand for higher data rates and quality of service (QoS);

– packet switch optimized system;

– continued demand for cost reduction;

– low complexity;

– avoid unnecessary fragmentation of technologies for paired and unpaired band operation.

In September 2009, the 3GPP partners made a formal submission to the International Telecommunication Union (ITU) proposing that LTE Release 10 and beyond (LTE Advanced) be evaluated as a candidate for IMT-Advanced. The ITU has coined the term International Mobile Telecommunication (IMT)-Advanced to identify mobile systems whose capabilities go beyond those of IMT-2000. In order to meet this new challenge, 3GPP's organizational partners agreed to widen 3GPP's scope to include the development of systems beyond 3G. Some of the key features of IMT-Advanced are:

– worldwide functionality and roaming;

– compatibility of services;

– interworking with other radio access systems;

– enhanced peak data rates to support advanced services and applications (100 Mbit/s for high and 1 Gbps for low mobility).

3.1. Relevant features of LTE

LTE is a mobile broadband solution that offers a rich set of features with a lot of flexibility in terms of deployment options and potential service offerings. Some of the most important features are as follows (Table 3.1):

– Orthogonal frequency-division multiplexing (OFDM) for high spectral efficiency is the basis of the physical layer: OFDM is used in downlink for robustness against multipath interference and high affinity to advanced techniques such as frequency domain channel-dependent scheduling and MIMO, while Single-Carrier Frequency Division Multiple Access (SC-FDMA) is used in uplink in order to get a low peak-to-average power ratio, user orthogonality in frequency domain and multi-antenna application.

– Support for TDD and FDD: LTE supports both time division duplexing (TDD) and frequency division duplexing. TDD is favored by a majority of implementations because of its advantages: (1) flexibility in choosing uplink-to-downlink data rate ratios, (2) ability to exploit channel reciprocity, (3) ability to implement in non-paired spectrum and (4) less complex transceiver design.

– Adaptive Modulation and Coding (AMC): LTE supports a number of modulation and forward error correction coding schemes and allows the scheme to be changed on a per user and frame basis, based on channel conditions. AMC is an effective mechanism to maximize throughput in a time-varying channel. The adaptation algorithm typically calls for the use of the highest modulation and coding scheme that can be supported by the signal-to-noise and interference ratio at the receiver such that each user is provided with the highest possible data rate that can be supported in their respective links.

– Support of variable bandwidths: E-UTRAN will operate in spectrum allocations of different sizes, including 1.25, 1.6, 2.5, 5, 10, 15 and 20 MHz in both the uplink and downlink (Table 1.3). Operation in the paired and unpaired spectrum will be supported. This scaling may be done dynamically to support user roaming across different networks that may have different bandwidth allocations.

– LTE is capable of supporting very high peak data rates. In fact, the peak physical data rate can be as high as the downlink peak data rate of 100 Mb/s within a 20 MHz downlink spectrum allocation (5 bps/Hz), while it provides an uplink peak data rate of 50 Mb/s (2.5 bps/Hz) within a 20 MHz uplink spectrum allocation.

– Mobility: E-UTRAN should be optimized for low mobile speed from 0 to 15 km/h. A higher mobile speed between 15 and 120 km/h should be supported with high performance. Mobility across the cellular network will be maintained at speeds from 120 to 350 km/h (or even up to 500 km/h depending on the frequency band).

– Link layer retransmissions: LTE supports automatic retransmission requests (ARQ) at the link layer. ARQ-enabled connections require each packet transmitted to be acknowledged by the receiver; unacknowledged packets are assumed to be lost and are retransmitted. LTE also optionally supports hybrid-ARQ, which is an effective hybrid between foreword error correction and ARQ.

– Simultaneous user support: LTE provides the ability to perform two-dimensional resource scheduling (in time and frequency), allowing the support of multiple users in a time slot; in contrast, existing 3G technology performs one-dimensional scheduling, which limits service to one user for each time slot. This capability of LTE results in a much better always-on experience and also enables the proliferation of embedded wireless applications/systems.

– Security: LTE provides enhanced security through the implementation of Universal Integrated Circuit Card Subscriber Identity Module (SIM) and the associated robust and non-invasive key storage and symmetric key authentication using 128-bit private keys. LTE additionally incorporates strong mutual authentication, user identity confidentiality, integrity protection of all signaling messages between user equipment and mobility management entity (MME) and optional multi-level bearer data encryption.

– Efficient worldwide roaming: because LTE is the unified pre-4G standard for most 3GPP and 3GPP2 carriers worldwide, LTE devices are fundamentally easier to set up for worldwide roaming. The caveat is that the actual frequency band used by different carriers is different (thereby retaining the need for multiband devices). As a result, the wireless migration path to LTE provides greater opportunities for seamless international roaming and for global device economies of scale. Table 3.1 depicts LTE Release 8 major parameters.

Parameter	*Value*
Peak data rate downlink DL	300 Mbps
Peak data rate uplink UL	75 Mbps
Transmission bandwidth DL	20 MHz
Transmission bandwidth UL	20 MHz
Mobility	Optimized for low speeds (<15 km/h), high performance at speeds up to 120 km/h and maintains links at speeds up to 350 km/h
Coverage	Full performance up to 5 km
Scalable bandwidths	1.4, 3, 5, 10, 15 and 20 MHz

Table 3.1. *LTE characteristics*

3.2. Network architecture and protocols

The 3GPP Long-Term Evolution/System Architecture Evolution (LTE/SAE) sought to take mobile technology to the next level through higher bandwidths, better spectrum efficiency, wider coverage and full interworking with other access/backend systems.

LTE/SAE proposed to do all this using an all-IP architecture with well-defined interworking with circuit-switched systems. The evolved 3GPP system introduced a hybrid mobile network architecture supporting radio access technologies and several mobility mechanisms. We begin this section by introducing the LTE network reference model and define its various functional entities and its interconnection possibilities. Next, we discuss the end-to-end protocol layering in an LTE network, network selection and discovery, and IP address allocation. Finally, we describe the functional architecture and processes associated with security, QoS and mobility management in more detail.

The network architecture of LTE is based on functional decomposition principles, where required features are decomposed into functional entities without specific implementation assumptions about physical network entities. This is why 3GPP specified a new packet core, the Evolved Packet Core (EPC), network architecture to support the E-UTRAN through a reduction in the number of network elements, simpler functionality, improved redundancy, and for connections and handover to other fixed line and wireless access technologies, giving service providers the ability to deliver a seamless experience.

3.2.1. *Architecture reference model*

Figure 3.1 shows the LTE network reference model, which is a logical representation of the network architecture. The network reference model identifies the functional entities in the architecture and the reference points between the functional entities over which interoperability is achieved. The overall architecture has two distinct components: the access network and the core network. The access network is E-UTRAN. The core network is all-IP core network and is fully packet switched. Services like voice, which are traditionally circuit switched, will be handled using IP Multimedia Subsystem (IMS) network. The core network is the EPC. Network complexity and latency are reduced as there are fewer hops in both the signaling and data plane. The EPC is designed to support non-3GPPP access supports for mobile IP. To improve system security, integrity protection and ciphering have been added via a non-access stratum plane, which is an additional layer of abstraction to protect important information like security interworking between 3GPP and non-3GPP network. Apart from the network entities handling data traffic, EPC also contains network control entities for keeping user subscription information represented by home subscriber server, determining the identity and privileges of a user and tracking his/her activities, i.e. authorization, authentication and accounting server, and enforcing charging and QoS policies through a Policy and Charging Rules Function (PCRF). Note that E-UTRAN and EPC together constitute the evolved packet system. Both radio access network and core network could achieve many functionalities including:

– network access control functions;

– packet routing and transfer functions;

– mobility management functions;

– security functions;

– radio resource management functions;

– network management functions.

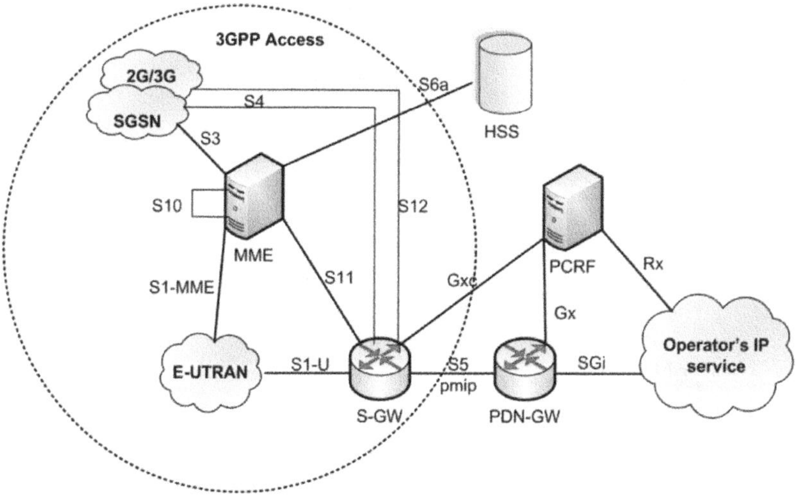

Figure 3.1. *LTE architecture. SESN – Serving GPRS support Node*

3.2.2. *Functional description of a LTE network*

3.2.2.1. *Evolved Universal Terrestrial Radio Access Network*

E-UTRAN is the air interface of 3GPP's LTE upgrade path for mobile networks. It is a radio access network standard meant to be a replacement of the Universal Mobile Telecommunications System, high-speed downlink packet access and high-speed uplink packet access technologies specified in 3GPP releases 5 and beyond. LTE's E-UTRAN is an entirely new air interface system, which provides higher data rates and lower latency and is optimized for packet data. It uses OFDMA radio access for the downlink and SC-FDMA for the uplink. The E-UTRAN in LTE architecture consists of a single node, i.e. the eNodeB that interfaces with the user equipment (UE). The aim of this simplification is to reduce the latency of all radio interface operations. eNodeBs are connected to each other via the X2 interface, and they connect to the packet switched core network via the S1 interface (see Figure 3.2).

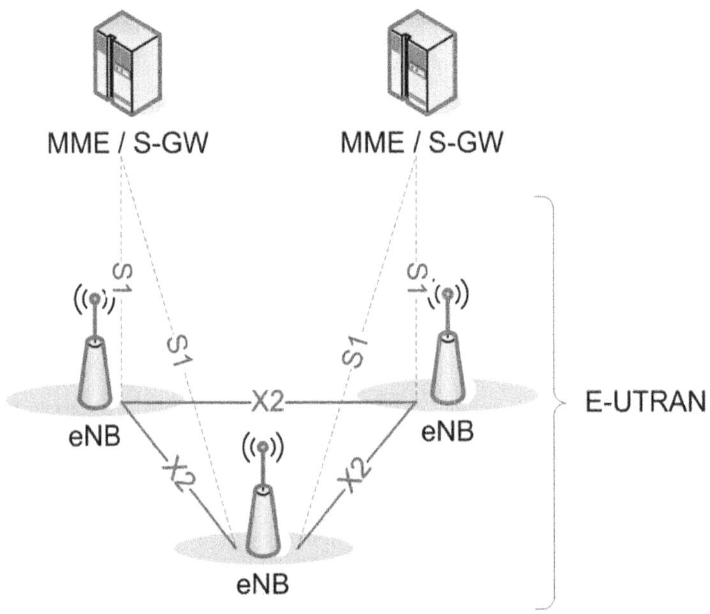

Figure 3.2. *E-UTRAN architecture. MME – mobility management entity; SGW – Serving gateway*

A general E-UTRAN protocol architecture (Figure 3.3) splits the radio interface into three layers: a physical layer or layer 1, the data link layer (layer 2) and the network layer or layer 3. This hierarchical stratification provides a complete vision of the radio interface, from both the functionality associated with each of the structured layer to the protocol flow between them. The purpose of the protocol stack is to set the services to organize the information to transmit through logical channels whose classifying parameter is the nature of the information they carry (i.e. control or traffic information) and map these logical channels into transport channels whose characteristic is how and with what the information within each logical channel is transmitted over the radio interface. This how and with what characteristic means that for each transport channel, there is one or more associated transport formats, each defined by the encoding, interleaving bit rate and mapping onto the physical channel. Each layer is characterized by the services provided to the higher layers or entities and the functions that support them as follows:

– *physical layer*: carries all information from the medium access channel (MAC) transport channels over the air interface. Takes care of the link adaptation (adoptive modulation and code (AMC)), power control, cell search (for initial synchronization and handover purposes) and other measurements (inside the LTE system and between systems) for the radiolink control (RRC) layer;

– *MAC*: the MAC sublayer offers a set of logical channels to the RLC sublayer that it multiplexes into the physical layer transport channels. It also manages the hybrid automatic repeat request (HARQ) error correction, handles the prioritization of the logical channels for the same UE and the dynamic scheduling between UEs, etc.

– RLC: it transports the Packet Data Convergence Protocols (PDCP's) PDUs. It can work in three different modes depending on the reliability provided. Depending on this mode, it can provide HARQ error correction, segmentation/ concatenation of PDUs, reordering for insequence delivery, duplicate detection, etc.

– PDCP: for the RRC layer, it transports its data with ciphering and integrity protection and for the IP layer it transports IP packets, with robust header compression (ROHC), ciphering and depending on the RLC mode in-sequence delivery, duplicate detection and retransmission of its own service data units during handover.

– RRC: between others, it takes care of the broadcast system information related to the access stratum and transport of the non-access stratum manages messages, paging, establishment and release of the radio resource control (RRC) connection, security key management, handover, UE measurements related to inter-system (inter-RAT) mobility, QoS, etc.

Layers interfacing, with the E-UTRAN protocol stack are:

– Non-access stratum: protocol between the UE and the mobility management entity (MME) on the network side (outside of EUTRAN). Between others it authenticates the UE and security control and generates part of the paging messages;

- IP layer.

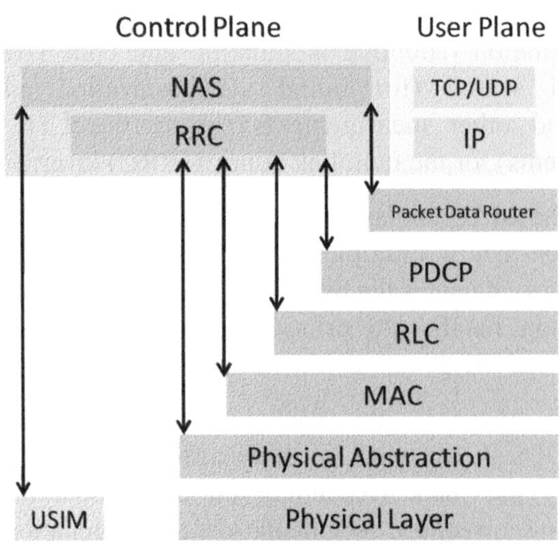

Figure 3.3. *LTE protocol layers*

3.2.3. *System architecture evolution*

The main component of the SAE architecture is the EPC which consists of the following functional elements:

– *Serving gateway (S-GW)*: the S-GW routes and forwards user data packets, while also acting as the mobility anchor for the user plane during inter-eNodeB handovers and as the anchor for mobility between LTE and other 3GPP technologies (terminating S4 interface and relaying the traffic between 2G/3G systems and the packet data network gateway (PDN-GW)). For idle state UEs, the S-GW terminates the downlink data path and triggers paging when downlink data arrives for the UE. It manages and stores UE contexts, e.g. parameters of the IP bearer service and network internal routing information. It also performs replication of the user traffic for lawful interception.

– *MME*: is the key control node for the LTE access network. It is responsible for idle mode UE tracking and paging procedures including retransmissions. It is involved in the bearer activation/deactivation process and is also responsible for choosing the S-GW for a UE at the initial attachment and at time of intra- LTE handover involving core network node relocation. It is responsible for authenticating the user. The non-access stratum signaling terminates at the MME and it is also responsible for the generation and allocation of temporary identities to UEs. It checks the authorization of the UE to camp on the service provider's Public Land Mobile Network (PLMN) and enforces UE roaming restrictions. The MME is the termination point in the network for ciphering/integrity protection for non-access stratum signaling and handles the security key management. Lawful interception of signaling is also supported by the MME. The MME also provides the control plane function for mobility between LTE and 2G/3G access networks with the S3 interface terminating at the MME from the SGSN.

Finally, the MME also terminates the S6a interface toward the home HSS for roaming UEs.

– *PDN-GW*: the PDN-GW provides connectivity to the UE to external packet data networks by being the point of exit and entry of UE traffic. A UE may have simultaneous connectivity with more than one PDN-GW for accessing multiple packet data networks. The

PDN-GW performs policy enforcement, packet filtering for each user, charging support, lawful interception and packet screening. Another key role of the PDN-GW is to act as the anchor for mobility between 3GPP and non-3GPP technologies such as WiMAX and 3GPP2 (CDMA 1x and EV-DO).

Figure 3.4 shows the functional split between E-UTRAN and EPC. Some functions after the split are indicated in this figure.

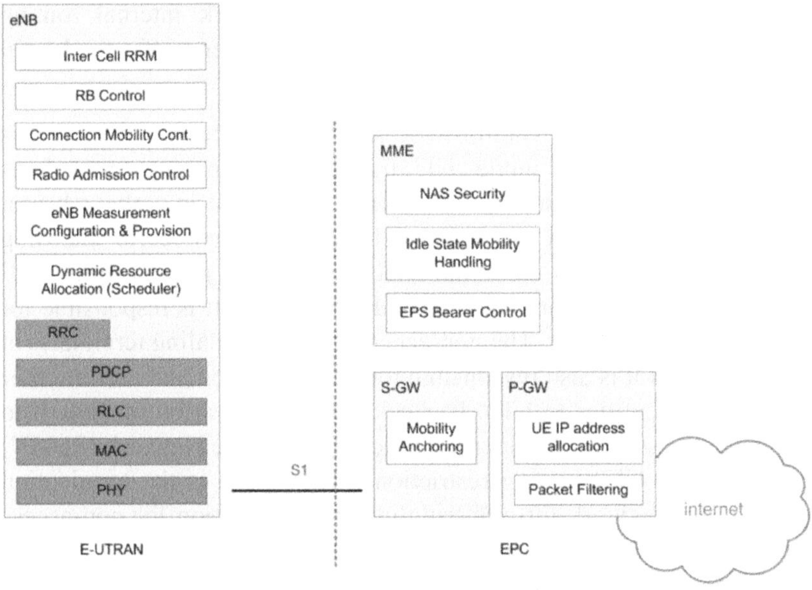

Figure 3.4. *Functional split between E-UTRAN and EPC*

3.2.4. *Reference points*

The LTE defines a reference point as a conceptual link that connects two groups of functions that reside in different functional entities of the E-UTRAN and EPC. Figure 3.5 shows a number of reference points defined by the 3GPP. These reference points are listed in Table 3.2. Note that these reference points are based on release 8 of the standardization and there may be more reference points that are dependent on the type of network architecture.

EPS entity name	*Function*
eNodeB	Radio resource management IP header compression and encryption of user data stream Selection of an MME at UE attachment when no routing to an MME can be determined Routing of user plane data toward the serving gateway Scheduling and transmission of paging messages Scheduling and transmission of broadcast information and measurement and measurement reporting Scheduling and transmission of PWS messages
MME	NAS signaling NAS signaling security AS security control Inter-CN node signaling for mobility between 3GPP access networks Idle mode UE reachability Tracking area list management (for UE in idle and active modes) PDN-GW and serving GW selection MME selection for handovers with MME change SGSN selection for handovers to 2G or 3G 3GPP access networks Roaming Authentication Bearer management functions including dedicated bearer establishment Support for PWS message transmission
S-GW	The local mobility anchor point for inter-eNodeB handover Mobility anchoring for inter-3GPP mobility E-UTRAN idle mode downlink packet buffering and initiation of network-triggered service request procedure Lawful interception Packet routing and forwarding Transport level packet marking in the uplink and the downlink Accounting on user and QCI granularity for interoperator charging UL and DL charging per UE, PDN, and QCI
PDN-GW	Per-user-based packet filtering Lawful interception UE IP address allocation Transport-level packet marking in the downlink UL and DL service-level charging, gating, and rate enforcement

Table 3.2. *Functional decomposition of the EPS*

3.3. Control and user planes

The radio interface in LTE is characterized through its protocols where it can be defined by two main groupings according to the final purpose: user plane protocols and control plane protocols. The first carries user data through the access stratum and the second is responsible for controlling the connections between the UE and the network and the radio access bearers. Even though separation of the control plane and the user plane was one of the most important issues of LTE design, full independence of the layers is not feasible because, without interaction between the user plane and the control plane, operators are not able to control QoS, the source/destination of media traffic or when the media starts and stops.

3.3.1. *User plane*

Figure 3.5 shows the user plane protocol stack including the E-UTRAN and the S1 interface of a conventional, i.e. non-self-backhauled, system. The radio access uses the MAC, RLC and PDCP protocols. The user plane part of the S1 interface is based on the GPRS Tunneling Protocol (GTP), which uses a tunneling mechanism ensuring that IP packets destined for a given UE are delivered to the eNodeB where the UE is currently located. GTP encapsulates the original IP packet into an outer IP packet which is addressed to the proper eNodeB. The S1 interface can operate over various layer 1/layer 2 technologies, e.g. fiber optic cables, leased (copper) lines or microwave links.

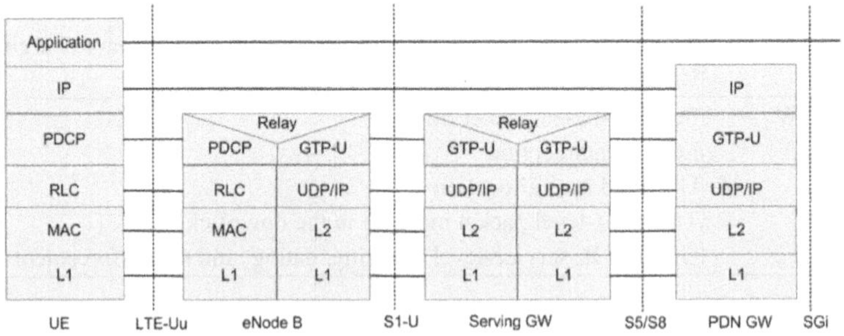

Figure 3.5. *User plane end-to-end protocol stack*

Figure 3.5 also shows an example of a Transmission Control Protocol (TCP)/IP-based application, such as web browsing. The corresponding peer entities operate in the UE and at the server hosting the web application. For simplicity, peer protocol entities of the server are drawn in the S-GW; however, in general, they are located somewhere in the Internet.

S1-U	E-UTRAN and S-GW	For per-bearer user plane tunneling and inter-eNodeB path switching during handover
S3	MME and SGSN	It enables user and bearer information exchange for inter-3GPP access network mobility in idle and/or active state
S4	S-GW and SGSN	It provides related control and mobility support between GPRS core and the 3GPP anchor function of S-GW
S5	S-GW and PDN-GW	It is used for S-GW relocation due to UE mobility and if the S-GW needs to connect to a non-collocated PDN-GW for the required PDN connectivity
S6a	MME and HSS	It enables transfer of subscription and authentication data for authenticating and authorizing user access between MME and HSS
S10	MME and MME	For MME relocation and MME to MME information transfer
S11	MME and S-GW	
S12	UTRAN and S-GW	For user plane tunneling when direct tunnel is established
Gx	PCRF and PDN-GW	It provides transfer of QoS policy and charging rules to Policy and Charging Enforcement Function (PCEF) in the PDN-GW
SGi	PDN-GW and PDN	PDN may be an operator – external public or private packet data network or an intra-operator packet data network, e.g. for provision of IMS services
Rx	PCRF and PDN	The Rx reference point resides between the AF and the PCRF

Table 3.3. *LTE reference points*

All information sent and received by the UE, such as the coded voice in a voice call or the packets in an Internet connection, are transported via the user plane. User plane traffic is processed at different hierarchical levels, from eNodeB up to the core network (EPC). Also, control traffic is strictly tied to the user plane.

Irrespective of the reasons behind the current hierarchical architecture, for the transmission backbone, this means the higher the level of network hierarchy, the greater the amount of accumulated traffic generated. Therefore, higher level network elements will readily become the bottleneck of the network. Therefore, transmission capacity should be fitted to the network hierarchy; at higher levels, high-capacity transmission means, such as fiber optics, are needed, but when it comes to the edge of the network, microwave transmission becomes a more flexible and cost-effective substitution, particularly in terms of extending capacity.

3.3.2. *GPRS tunneling protocol*

GPRS Tunneling Protocol (GTP) is a collection of protocols central to IP mobility management within 3GPP packet core networks (GPRS/UMTS/EPC) comprising of GTP-C, GTP-U and GTP variants. The protocol stack for GTP is depicted in Figure 3.6. GTP-C is the control part of GTP and is used in control plane mechanisms in GPRS, UMTS and LTE/SAE/EPC networks. GTP-C is standardized as version 0, version 1 and version 2 by 3GPP. All the GTP-C versions use UDP as the transport protocol. GTP v2 offers a fallback to GTP v1 via the earlier "Version Not Supported" mechanism but explicitly offers no support for fallback to GTP v0.

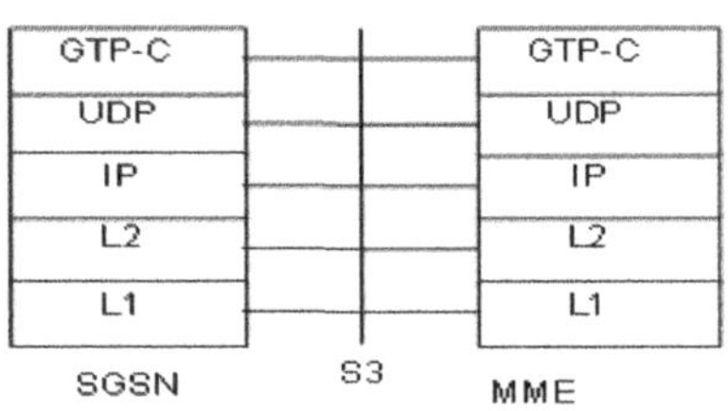

Figure 3.6. *GTP stack*

GTP-U is the bearer part of GTP and is used in user plane mechanisms in GPRS, UMTS and LTE networks. GTP-U is standardized as version 0 and version 1 by 3GPP. All the GTP-U versions use UDP as the transport protocol. GTP or GTP Prime is used for interfacing with the changing gateway function (CGF) in

GPRS and UMTS networks. LTE MME, SGW and PDN gateway nodes use GTP-C to control plane signaling on S11/S5 interfaces, while S-GW and PDN-GW nodes use GTP-U as a user plane on S1-U and S5 interfaces primarily. The LTE/SAE/EPC network uses only GTP version 2, also known as evolved GTP unless backward compatible to 3G UMTS/HSPA networks. After the downlink path is switched at the S-GW downlink, packets on the forwarding path and on the new direct path may arrive interchanged at target eNodeB. Target eNodeB should first deliver all forwarded packets to the UE before sending any of the packets received on the new direct path. The method employed in the target eNodeB to enforce the correct delivery order of packets is outside the scope of the standard. In order to assist the reordering function in the target eNodeB, the S-GW will send one or more "end marker" packets on the old path immediately after switching the path for each UE. The "end marker" packet will not contain user data. The "end marker" is indicated in the GTP header. After sending the tagged packets, the gateway will not send any further user data packets via the old path. Upon receiving the "end marker" packets, the source eNodeB will, if forwarding is activated for that bearer, forward the packet toward the target eNodeB. On detection of an "end marker," the target eNodeB will discard the end marker packet and initiate any necessary processing to maintain in-sequence delivery of user data forwarded over the X2 interface and user data received from the S-GW over S1 as a result of the path switch. On detection of the "end marker," the target eNodeB may also initiate release of the data forwarding resource (see Figure 3.7).

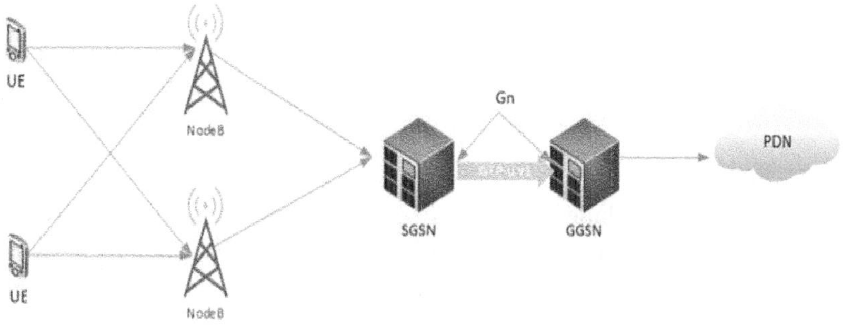

Figure 3.7. *GTP tunneling*

3.3.3. *Control plane*

The function of the control plane protocol is to control the radio access bearers and the connection between the UE and the network, i.e. signaling between E-UTRAN and EPC (Figure 3.8). The control plane consists of protocols for control and support of the user plane functions:

– controlling the E-UTRAN network access connections, such as attaching to and detaching from E-UTRAN;

– controlling the attributes of an established network access connection, such as activation of an IP address;

– controlling the routing path of an established network connection in order to support user mobility;

– controlling the assignment of network resources to meet changing user demands.

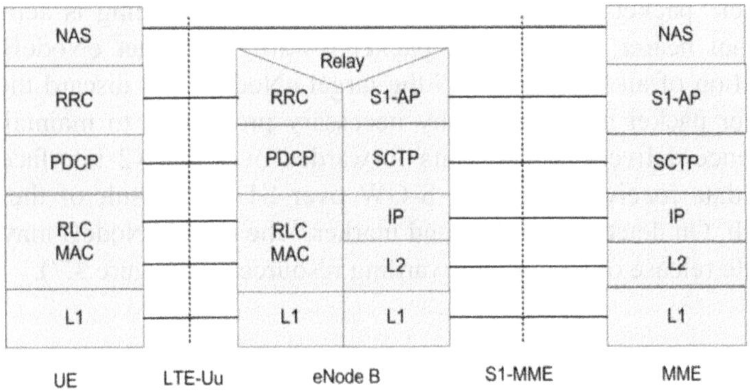

Figure 3.8. *Control plane end-to-end protocol stack*

In the control plane, the non-access stratum (NAS) protocol, which runs the MME and the UE, is used for control actions such as network attach, authentication, setting up of bearers and mobility management. All NAS messages are ciphered and integrity is protected by the MME and UE. The RRC layer in eNodeB makes handover decisions based on neighbor cell measurements sent by the UE, pages for the UEs over the air, broadcasts system information, controls UE measurement reporting such as the periodicity of channel quality information, reports on and allocates temporary cell-level identifiers to active UEs. It also executes transfer of UE context from the source eNodeB to the target eNodeB during handover and does integrity protection of RRC messages. The RRC layer is responsible for setting up and maintaining radio bearers.

3.4. Multimedia broadcast and multicast service

Multimedia broadcast and multicast service (MBMS) is a point-to-multipoint service in which data are transmitted from a single source to multiple destinations over radio network. Transmission of the same data to multiple recipients allows network resources to be shared. MBMS is realized by the addition of existing and new functional entities of 3GPP architecture.

MBMS provides two different services: (1) broadcast and (2) multicast. The broadcast service can be received by any subscriber located in the area in which the service is offered and multicast services can only be received by users having subscribed to the service who have joined the multicast group associated with the service. Both services are unidirectional point-to-multipoint transmissions of multimedia data and can be applied to text, audio, picture and, video broadcast from a Broadcast Multicast Service Center to any user located in the service area. For such a service, only broadcast service providers can be charged, possibly based on the amount of data broadcast, size of service area or broadcast duration. Multicast is subject to service subscription and requires the end user to explicitly join the group in order to receive the service. Because it is subject to subscription, the multicast service allows the operator to set specific user charging rules for this service.

3.5. Stream Control Transmission Protocol

The Stream Control Transmission Protocol (SCTP) is a transport layer protocol, serving a similar role to the popular TCP and UDP. It provides some of the same service features: it is message-oriented like UDP and ensures reliable, in-sequence transport of messages with congestion control like TCP. SCTP is used in LTE to ensure reliable, in-sequence transport of messages. LTE uses SCTP, which we view as a layer between the SCTP user application and an unreliable end-to-end datagram service such as UDP. Thus, the main function of SCTP amounts to reliable transfer of user datagrams between peer SCTP users. It performs this service within the context of an association between SCTP nodes, where APIs exist at the boundaries. SCTP has connection-oriented characteristics but a broad concept. It provides a means for each SCTP endpoint to provide the other during association startup with a list of transport addresses (e.g. address/UDP port combinations) by which that endpoint can be reached and from which messages originate. The association transfers over all possible source/destination combinations, which may be generated from two end lists. As a result, SCTP offers the following services:

– application-level segmentation;

– acknowledged error-free non-duplicated transfer of user data;

– sequenced delivery of user datagrams within multiple streams;

– enhanced reliability through support of multi-homing at either or both ends of the association;

– optional multiplexing of the user datagram into SCTP datagrams.

SCTP assumes that it is running over an IPv4 or IPv6 network. Even more importantly, it assumes it is running over a well-engineered IP network. This, in practice, means that there is a diverse routing network underneath to avoid a single point of failure. In LTE, SCTP handles communications between the eNodeB and the MME. This communication connection is very important and fragile since it must be able to detect dropouts very quickly. TCP does not do this, whereas SCTP immediately detects and recognizes when a packet is dropped or a link goes down. LTE providers specifically and telecom networks in general need this ability to ensure a high QoS. Additionally SCTP has, as a default, "selective ACK" which is optional in TCP. What this means is that a packet will never be resent if it has already been acknowledged as sent. In the LTE world, where every bit counts, using SCTP means no wasted data. The purpose of the use of SCTP in LTE is to provide a robust and reliable signaling bearer. To achieve this, SCTP provides appropriate congestion control procedures, fast retransmitssion in the case of message loss, and enhanced reliability. It also provides additional security against blind attacks and will be used to increase security in connecting the LTE networks of different operators.

3.6. Network discovery and selection

The Dynamic Host Control Protocol (DHCP) is the primary mechanism for allocating a dynamic Point-of-Attachment (PoA) IP address to the UE. Note that the EPS bearer supports dual-stack IP addressing, meaning that it is able to transport both native IPv4 and native IPv6 packets. In order to support DHCP-based IP address configuration in (IPv4 and IPv6), the PDN-GW will act as the DHCP server for Home Public Land Mobile Network (HPLMN)-assigned dynamic and static and Visited Public Land Mobile Network (VPLMN)-assigned dynamic IP addressing. When DHCP is used for external PDN-assigned addressing and parameter configuration, the PDN-GW will act as the DHCP server toward the UE and it will act as the DHCP client toward the external DHCP server. The serving gateway does not have any DHCP functionality. It forwards all packets to and from the UE including DHCP packets as normal.

The IPv6 Stateless Address auto configuration is the basic mechanism to allocate a /64 IPv6 prefix to the UE. Alternatively shorter than /64 IPv6 prefix delegation via DHCPv6, RFC 3633 may be provided, if it is supported by the PDN-GW. When DHCPv6 prefix delegation is not supported, the UE should use stateless address auto configuration RFC 4862.

3.7. Radio resource management

The purpose of radio resource management (RRM) is to ensure efficient use of the radio resources available and to provide mechanisms that enable E-UTRAN to meet radio resource-related requirements like: (1) enhanced support for end-to-end QoS, (2) efficient support for transmission of higher layers and (3) support of load sharing and policy management across different radio access technologies. RRM in E-UTRAN provides a means to manage (e.g. assign, re-assign and release) radio resources taking into account single- and multi-cell aspects. The RRM functions are represented by:

– *Radio Bearer Control (RBC)*: The establishment, maintenance and release of radio bearers involves the configuration of associated radio resources. When setting up a radio bearer for a service, RBC takes into account the overall resource situation in E-UTRAN, the QoS requirements of in-progress sessions and the QoS requirement for the new service. RBC is also concerned with the maintenance of radio bearers of in-progress sessions when the radio resource situation changes due to mobility or other reasons. RBC is involved in the release of radio resources associated with radio bearers at session termination, handover or on other occasions. RBC is located in the eNodeB.

– *Connection Mobility Control (CMC)*: CMC is concerned with the management of radio resources in connection with idle or connected mode mobility. In idle mode, the cell reselection algorithms are controlled by setting parameters (thresholds and hysteresis values) that define the best cell and/or determine when the UE should select a new cell. E-UTRAN broadcasts parameters that configure the UE measurement and reporting procedures. In connected mode, the mobility of radio connections has to be supported. Handover decisions may be based on UE and eNodeB measurements. In addition, handover decisions may take other inputs, such as neighbor cell load, traffic distribution, transport and hardware resources, and operator-defined policies into account. CMC is located in the eNodeB.

– *Dynamic Resource Allocation (DRA) or Packet Scheduling (PS)* allocates and deallocates resources (including buffer and processing resources and resource blocks [i.e. chunks]) to user and control plane packets. DRA involves several subtasks, including the selection of radio bearers whose packets are to be scheduled and manages the necessary resources (e.g. the power levels or the specific resource blocks used). PS typically takes into account the QoS requirements associated with the radio bearers, the channel quality information for UEs, buffer status, interference situation, etc. DRA may also take into account restrictions or preferences on some of the available resource blocks or resource block sets due to intercell interference coordination (ICIC) considerations. DRA is located in the eNodeB.

– *intercell interference coordination (ICIC)* manages radio resources (notably, radio resource blocks) such that intercell interference is kept under control. ICIC is inherently a multi-cell RRM function that needs to take into account information (e.g. the resource usage status and traffic load situation) from multiple cells. The preferred ICIC method may be different in the uplink and downlink. ICIC is located in the eNodeB.

– *Load Balancing (LB)* has the task to handle the uneven distribution of traffic over multiple cells. The purpose of LB is thus to influence the load distribution in such a manner that radio resources remain highly utilized and the QoS of for sessions in progress is maintained to the best possible extent and call dropping probabilities are kept sufficiently small. LB algorithms may result in handover or cell reselection decisions with the purpose of redistributing traffic from highly loaded cells to underutilized cells. LB is located in the eNodeB.

– *Inter-RAT RRM* is primarily concerned with the management of radio resources in connection with inter-RAT mobility, notably inter-RAT handover. At inter-RAT handover, the handover decision may take into account the RAT resource situation as well as UE capabilities and operator policies. The importance of inter-RAT RRM may depend on the specific scenario in which E-UTRAN is deployed. Inter-RAT RRM may also include functionality for inter-RAT load balancing for idle and connected mode UEs.

– *Subscriber profile ID for RAT/frequency priority (SPID)* The RRM strategy in E-UTRAN may be based on user-specific information. The SPID parameter received by the eNodeB via the S1 interface is an index referring to user information (e.g. mobility profile and service usage profile). The information is UE specific and applies to all its radio bearers. This index is mapped by the eNodeB to a locally defined configuration in order to apply specific RRM strategies (e.g. to define RRC_IDLE mode priorities and control inter-RAT/interfrequency handover in RRC_CONNECTED mode).

3.8. Authentication and authorization

The trust model in LTE (see Figure 3.9) is similar to that of UMTS. It can roughly be described as a secure core network, while radio access nodes and interfaces between the core network and the radio access nodes are vulnerable to attack. The system architecture for LTE is flatter than that of UMTS, having no node that corresponds to the Radio Network Controller (RNC) in UMTS. Therefore, the UE user plane security must be terminated either in the LTE eNodeB or in a core network node.

For reasons of efficiency, it has been terminated in the eNodeB. However, because eNodeBs and backhaul links might be deployed in locations that are vulnerable to attack, some new security mechanisms have been added. Security over the LTE air interface is provided through strong cryptographic techniques. The backhaul link from the eNodeB to the core network makes use of Internet Key Exchange (IKE) and the IP Security Protocol (IPSec) when cryptographic protection is needed. Strong cryptographic techniques provide end-to-end protection for signaling between the core network and UE. Therefore, the main location where user traffic is threatened by exposure is in the eNodeB. To minimize susceptibility to attacks, the eNodeB needs to provide a secure environment that supports the execution of sensitive operations, such as the encryption or decryption of user data and the storage of sensitive data like keys for securing UE communication, long-term cryptographic secrets and vital configuration data. Likewise, the use of sensitive data must be confined to this secure environment. Even with the above security measures in place, one must consider attacks on an eNodeB, because, if successful, they could give attackers full control of eNodeB and its signaling to UEs and other nodes. To limit the effect of a successful attack on one eNodeB, attackers must not be able to intercept or manipulate user and signaling plane traffic that traverses another eNodeB – for example, after handover.

3.8.1. *User authentication, key agreement and key generation*

The subscriber-authentication function in the LTE/3GPP EPS is based on the UMTS Authentication and Key Agreement (UMTS AKA) protocol. It provides mutual authentication between the UE and core network, ensuring robust charging and guaranteeing that no fraudulent entities can pose as a valid network node. Note that GSM SIMs are not allowed in LTE because they do not provide adequate security. EPS AKA provides a root key from which a key hierarchy is derived. The keys in this hierarchy are used to protect signaling and user plane traffic between the UE and network. The key hierarchy is derived using cryptographic functions. For example, if key2 and key3 (used in two different eNodeBs) are keys derived from key1 by a

MME, an attacker who gets hold of key2 still cannot deduce key3 or key1, which is on a higher layer in the key hierarchy.

Keys are bound to where, how and for which purpose they are used. This ensures, for example, that keys used for one access network cannot be used in another access network, and that the same key is not used for multiple purposes or with different algorithms. Because GSM does not have this feature, attackers who break one algorithm in GSM can also compromise the security offered when other algorithms use the same key. The key hierarchy and binding also make it possible to routinely and efficiently change the keys used between a UE and eNodeBs (for example, during handover) without changing the root key or the keys used to protect signaling between the UE and core network.

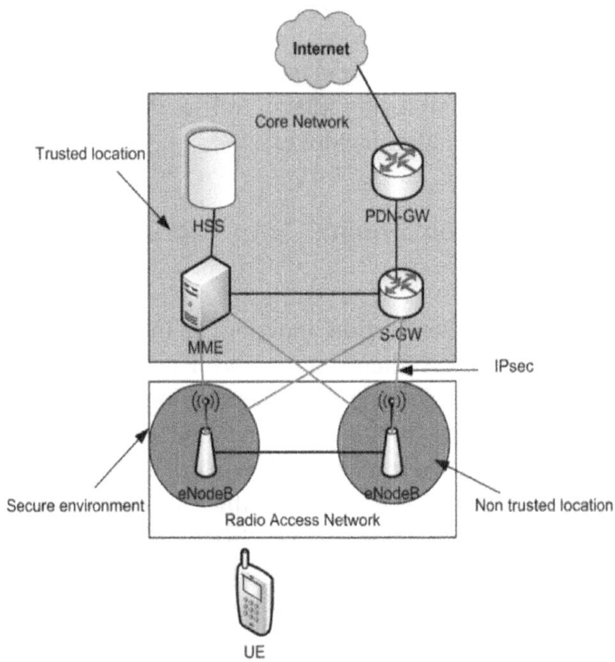

Figure 3.9. *LTE trusted model*

3.8.2. *Signaling and user-plane security*

For radio-specific signaling, LTE provides integrity, replay protection and encryption between the UE and eNodeB. Internet Key Exchange (IKE)/IPSec can protect the backhaul signaling between the eNodeB and MME. In addition, LTE-specific protocols provide end-to-end protection of signaling between the MME and UE. For user-plane traffic, IKE/IPSec can protect the backhaul from the eNodeB to the S-GW). Support for integrity, replay protection and encryption is mandatory in eNodeB. The user-plane traffic between the UE and eNodeB is only protected by encryption as integrity protection would result in expensive bandwidth overhead. It is not possible to intelligently inject traffic on behalf of another user: attackers are essentially blind in the sense that any traffic they try to inject would almost certainly decrypt to garbage.

3.9. Fundamentals of the MAC layer in LTE

The important feature of the MAC layer in LTE eNB is to offer RRC functionalities related to the control plane. It achieves different functions: RRM admission control, scheduling and enforcement of negotiated QoS, cell information broadcast, ciphering/deciphering of user, control plane data and compression/decompression of downlink/uplink user plane packet headers. The protocol architecture of the LTE air interface can be separated between control and user planes. In the user plane, the application creates data packets that are processed by protocols such as TCP, UDP and IP; instead, in the control plane, the RRC protocol generates the signaling messages that are exchanged between eNB and UE.

3.9.1. *Traffic classes and quality of service*

Managing the quality of user experience in LTE is an issue which is treated carefully by 3GPP as it introduced some QoS mechanisms that allow the operators to best manage the different quality of experience of users subject to different types of application use. Such QoS mechanisms are introduced in both control and use planes. For example, the control plane mechanisms are used to permit the negotiation between the UEs and the network in order to agree on the different QoS requirements, recognize which UEs and applications are associated with which type of QoS and let the network correctly assign resources to each type of service. However, the QoS mechanism in the user plane has the role of agreeing those QoS requirements and controlling network resource consumption for an application/user pair. This leads to a conclusion that many applications can be run in a UE and eNB in the same time, each of which has different QoS parameters and requirements. Consequently, it is imperative to set up different bearers to support those multiple QoS requirements. Each bearer is associated with a QoS support mechanism. Broadly speaking, bearers in LTE are categorized into classes depending on the nature of the QoS they are providing: (1) minimum guaranteed bit rate (GBR) bearers, which are used mainly for real-time applications that require a guaranteed QoS parameters and (2) non-GBR bearers associated with non-real-time applications that do not need a permanent resource allocation guarantee.

3.9.2. *Mobility*

LTE introduced a hybrid mobile network architecture supporting radio access technologies and several mobility mechanisms. The EPC introduced an important mobility management entity called MME, which manages mobility when UEs are handing over from one cell to another in a very transparent way to the end user. LTE defined the procedures and the protocols to achieve mobility and handover to/from 3GPP and non-3GPP technologies. In this chapter, we are focusing more on the case when there is handover between macro/femtocell and macro/picocells. We are interested in the handover of UEs among LTE-A technology based HetNet. The movement from macrocell to femtocell is quite difficult due to the large number of candidate femtocells in the area of convergence and the characteristics of the femtocell entity. These characteristics can be defined as the different access modes introduced in a femtocell and the radio resource allocation method. The selection and decision to handover to femtocell access point are hard to make. However, when users of macrocells are trying to handover to a picocell, other issues and challenges should be considered, this can be represented by the cell range expansion introduced in a picocell, the mobility is considered to be load balanced as by increasing the range of the cell, the probability of handover from macrocell to picocells will increase and it will not be based on the quality of signal but mainly on the load on the picocells.

3.9.3. *Resource scheduling algorithms*

In a mobile communication, the Robust Secure Routing Protocol (RSRP) of UEs won't be the same since this is due to different factors related mainly to the network environment, such as UE location. The best CQIs from the users always become the basis radio bearer selection in resource scheduling assignments. In other words, the scheduling operator selects the UEs whose channel quality is better than those located in the cell-edge. The cell-edge UEs normally experience more interference from the neighboring signal than the cell-center UEs. Although the best CQI can achieve a high system throughput, conversely those UEs who have received a poor signal, due to interference parameter or are far away from the served base station, as will not participate in the network scheduling system.

Selecting UEs with the best CQI for resource scheduling is an unfair problem in the resource assignment process. In this chapter, we have adopted two opportunistic scheduling strategies for scheduling the whole contribution:

– Exponential Proportional Fair (EX-PF) which is able to increase fairness among UEs. The scheduler scheme assigns the UEs by comparing the ratio of current data rate to the average data rate of a particular users. Fairness can be addressed in this strategy while maintaining comparable long-term throughput for all users. Along with fairness and throughput, packet delay is a key parameter in measuring the performance of high delay sensitive real-time applications like video streaming, etc., while the presented scheduling does not consider packet delay for each individual user. As a result, it is not suitable for the scheduling of real time services.

– Modified Largest Weighted Delay First (MLWDF) is able to handle delay-sensitive traffic well. It schedules the UEs by comparing the combination of packet delay, current data rate and average data rate in an optimal way. MLWDF exploits asynchronous channel variations and allows different mobile terminals to transmit at a different bit-rate or signal-to-interface-plus-noise-ratio (SINR), so that higher efficiency can be achieved. However, it does not provide explicit QoS guarantees such as data rate to cell-edge users.

3.10. Fundamentals of the LTE physical layer

The physical LTE layer is based on OFDMA technology. OFDMA is the multiple access/multiplexing transmission scheme which has high efficiency, multiplexing data stream operations from multiple users to enable high-speed data, video and multimedia communications and has been used by a variety of commercial broadband systems, including DSL, WiFi, etc. In this section, we cover the basics of OFDMA and provide an overview of the physical LTE layer.

3.10.1. *Slot and frame structure in LTE OFDMA*

Before detailing the physical layer's generic frame structure, it is worth mention in that the downlink and uplink transmissions in LTE use one of two duplexing modes: TDD or FDD in which the duplex is realized in time and frequency, respectively. TDD and FDD modes are widely deployed in the existing cellular system; they both have their own advantages and disadvantages. The operator usually selects the duplexing format according to the applications supported. However, in practical deployment systems, FDD is the most frequently used mode due to its reduced complexity.

OFDMA depends on an OFDM physical layer. OFDM is a multicarrier transmission technique that divides the available bandwidth into a large number of small radio-frequency channels. In other words, OFDM is a multiplexing technique that subdivides the bandwidth into multiple frequency subcarriers, each set subcarriers is sub-channel or resource block (RB). An RB is a grid of rectangular resource elements, which consists of 12 consecutive subcarriers in the frequency domain and six or seven OFDMA symbols in the time domain. Thus, in an OFDM system, a RB is available in the time domain by means of OFDM symbols and in the frequency domain by means of subcarriers. The total number of RBs available depends on the overall transmission bandwidth of the system. LTE-A specification defines different exploitable bandwidths from 1.4 to 20 MHz which correspond, to 6 and 100 RBs respectively.

Generally, in a LTE OFDM physical system, the input data stream is divided into several parallel substreams of reduced data rate and each substream is modulated and transmitted on separate orthogonal subcarriers, while the orthogonality among carriers depends on the cyclic prefix (CP) duration or the increased symbol duration improve the robustness of OFDM to delay spread. It means that the CP can completely eliminate inter-symbol interference when CP duration is longer than the channel delay spread. The minimum allocation bandwidth for a UE is one RB (180 KHz), which is carried out by the MAC scheme in each subframe as it is the sublayer which is responsible for scheduling transmission over the LTE air interface in both downlink/uplink directions. It must be noted that in all OFDMA systems, the scheduler decides which users are allowed to transmit and which RBs to assign to each user. Usually, the RB assignment to UEs operates under each 1 ms transmission time interval. The scheduler must also decide the quantity of power to apply to each RB as well as a suitable modulation and coding scheme (MCS) to be assigned to each user. Figure 3.10 shows the LTE-A radio frame structure of 10 ms duration divided into 10 subframes, each consisting of two slots of 0.5 ms.

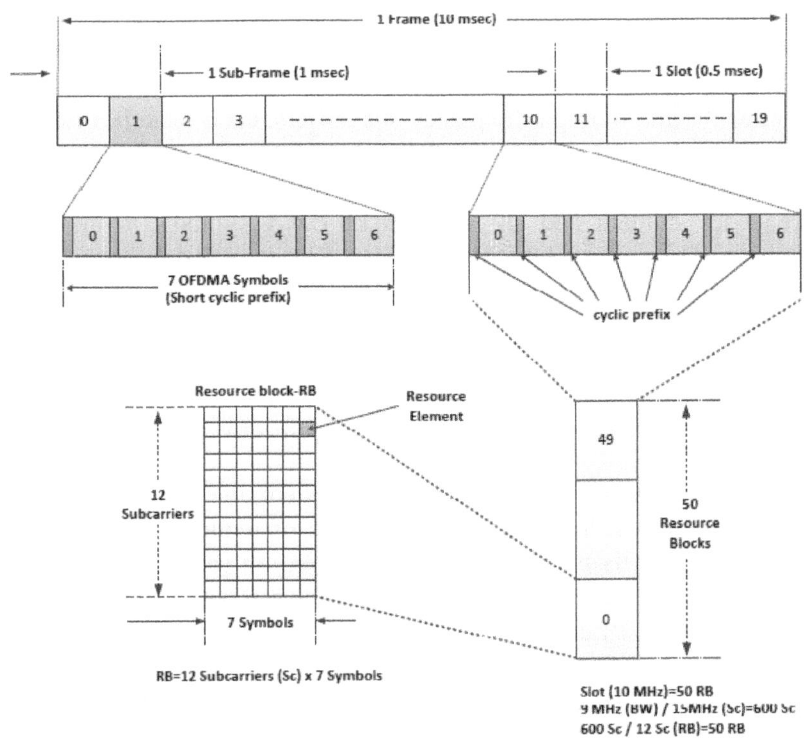

Figure 3.10. *Generic frame structure in the LTE/E-UTRA downlink*

Due to frequency-selective interference and fading fluctuations along the spectrum, different users can experience diverse channel conditions in a given subcarrier. While the contiguous permutation group subcarriers form RB, which is also known as band-AMC, it enables multiuser diversity by choosing subcarriers for the mobile terminal with the best signal quality received. In the OFDMA system, the transmitters need the average CQI feedback from users to apply the link adaptation, precoding preequalization, and adaptive transmission antenna diversity in order to maximize spectral efficiency. Table 3.3 describes the different MCSs supported by LTE-A, indexed according to CQI. As explained, if the channel is high fade, low order modulation is employed; conversely, if the channel is in good condition a high data rate can be achieved by employing high-order modulation. Similarly, code rate and transmission power can be optimized according to the instantaneous channel conditions and the required transmission rate and reliability. Moreover, when a mobile terminal is assigned with diversity subchannelization, it sends the reported CQIs back to its base station. This is equivalent to the SINR level over each of the divided subcarriers or RBs. According to the CQI method reported, the base station can know the SINR, and then determine the appropriate MCS level for each RB. These options enable the system designer to trade mobility for throughput.

MCS	Modulation	Code Rate	SINR threshold [dB]	Spectral Efficiency [bits/symbol/Hz]
MCS1	QPSK	1/12	-6.50	0.15
MCS2	QPSK	1/9	-4.00	0.23
MCS3	QPSK	1/6	-2.60	0.38
MCS4	QPSK	1/3	-1.00	0.60
MCS5	QPSK	1/2	1.00	0.88
MCS6	QPSK	3/5	3.00	1.18
MCS7	16QAM	1/3	6.60	1.48
MCS8	16QAM	1/2	10.00	1.91
MCS9	16QAM	3/5	11.40	2.41
MCS10	64QAM	1/2	11.80	2.73
MCS11	64QAM	1/2	13.00	3.32
MCS12	64QAM	3/5	13.80	3.90
MCS13	64QAM	3/4	15.60	4.52
MCS14	64QAM	5/6	16.80	5.12
MCS15	64QAM	11/12	17.60	5.55

Table 3.3. *Modulation and coding schemes for LTE (3GPP Release 10)*

3.10.2. *Reference signals*

To allow for coherent demodulation at the UE, reference symbols (or pilot symbols) are inserted in the OFDM time-frequency grid to allow for channel estimation. Downlink reference symbols are inserted within the first and third last OFDM symbol of each slot with a frequency domain spacing of six subcarriers (this corresponds to the fifth and fourth OFDM symbols of the slot for a normal and extended cyclic prefix, respectively) as shown in Figure 3.10 for LTE system with one antenna in normal CP mode. Furthermore, there is frequency domain staggering of three sub carriers between the first and second reference symbols. Therefore, there are four reference symbols within each RB. The UE will interpolate over multiple reference symbols to estimate the channel. For two transmission antennas, reference signals are inserted from each antenna where the reference signals on the second antenna are offset in the frequency domain by three sub carriers. To allow the UE to accurately estimate the channel coefficients, nothing is transmitted on the other antenna at the same time-frequency location of the reference signals.

The reference symbols have complex values, which are determined according to the symbol position as well as the cell. LTE specifications refer to this as a two-dimensional reference-signal sequence, which indicates the LTE cell identity. There are 510 reference signal sequences corresponding to 510 different cell identities. The reference signals are derived from the production of a two-dimensional pseudo-random sequence and a two-dimensional orthogonal sequence. There are 170 different pseudo-random sequences corresponding to 170 cell identity groups and three orthogonal sequences, each corresponding to a specific cell identity within the cell identity group. Reference signals are generated as the product of an orthogonal sequence and a pseudo-random numerical (PRN) sequence. Overall, there are 510 unique reference signals possible. A specified reference signal is assigned to each cell within a network and acts as a cell-specific identifier.

As shown in Figure 3.11, reference signals are transmitted on equally spaced subcarriers within the first and third last OFDM symbol of each slot. UE must get an accurate committed information rate (CIR) from each transmitting antenna. Therefore, when a reference signal is transmitted from one antenna port, the other antenna ports in the cell are idle. Reference signals are sent on every sixth subcarrier. CIR estimates for subcarriers that do not bear reference signals are computed via interpolation. Changing the subcarriers that bear reference signals by pseudo-random frequency hopping is also under consideration.

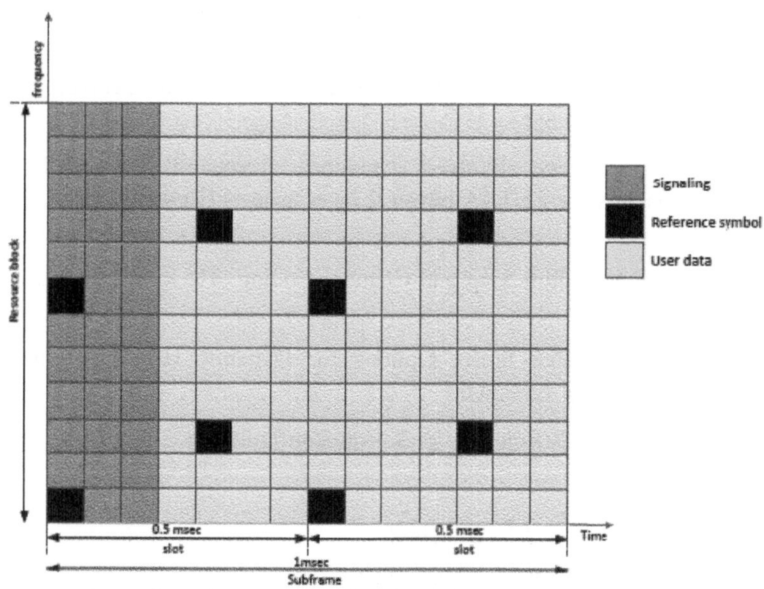

Figure 3.11. *Location of a reference symbol within a resource block*

3.11. Conclusion

The LTE generation is now well established in most countries. New improvements are included UMTS release 9, that introduces many extensions in femtocells, machine-to-machine, public safety warnings (Tsunami, etc.), and finally a green agenda (optimization of eNodeB). The latter was particularly exciting given the tremendous increase in the capacity requested by customers.

3.12. Bibliography

3.12.1. *Standards*

3GPP TR 25.913 Requirements for Evolved UTRA (E-UTRA) and Evolved UTRAN (EUTRAN).

3GPP TS 22.246 Multimedia Broadcast/Multicast Service (MBMS) User Services: Stage 1.

3GPP TS 24.301 Non-Access-Stratum (NAS) Protocol for Evolved Packet System (EPS): Stage 3.

3GPP TS 36 Access, Evolved Universal Terrestrial Radio. Further advancements for EUTRA physical layer aspects (Release 9): V9. 2, 13, 2010.

3GPP, E-UTRA Multiplexing and Channel Coding, TS 36.212, Release 12, 2013. 13.

IETF RFC 3633 IPv6 Prefix Options for Dynamic Host Configuration Protocol (DHCP) version 6.

IETF RFC 462 IPv6 Stateless Address Autoconfiguration.

3.12.2. *Selected bibliography*

[ALI 11] ALI-YAHIYA T., *LTE Femtocell Integration with Wireless Sensor/Actuator Networks and RFID Technologies*, Springer, 2011.

[DAH 10] DAHLMAN E., PARKVALL S., SKOLD J. *ET AL.*, *3G Evolution: HSPA and LTE for Mobile Broadband*, Academic Press, 2010.

[GIA 13] GIAMBENE G., ALI YAHIYA T., "LTE planning for soft frequency reuse", *Wireless Days*, IFIP, IEEE, pp. 1–7, 2013.

[HOB 09] HOBBY J.D., CLAUSSEN H., "Deployment options for femtocells and their impact on existing macrocellular networks", *Bell Labs Technical Journal,* vol. 13, no. 4, pp. 145–160, 2009.

[MOT XX] MOTOROLA, Long-term Evolution (LTE): a Technical Overview, Technical White Paper.

4

Long-Term Evolution Advanced

LTE Advanced (LTE–A) is a real broadband wireless network that provides peak data rates equal to or even greater than those for wired networks, i.e. Fiber To The Home (FTTH), while providing better quality of service (QoS). The major high-level requirements of LTE-A are reduced network cost (cost per bit), better service provisioning and compatibility with 3GPP systems. LTE-A is backward compatible with LTE. LTE-A was standardized by 3GPP UMTS Release 10 with the following main items: Fill native IP, Multiple MIMO, smart antenna, directive antenna, cognitive radio, relay and mesh techniques, green properties, and femtocells.

Extensions are provided with UMTS Release 11 with the introduction of service-oriented mobile networks, heterogeneous networks (HetNet), Coordinated Multi-Point (CoMP) operation and In-Device Co-Existence (IDC) allowing us to take into account interference between bands due to signal distortion.

More extensions are provided with Release 12:

– improvements in spectral efficiency;

– LTE carrier aggregation;

– multi access;

– machine-type communications (MTCs);

– interworking between mobile operators;

– continuity of data sessions in local networks;

– telepresence.

Some extensions are available through Release 13 which should increase focus on virtualization and LTE Unlicensed (LTE-U) allowing the aggregation of primary carriers in a licensed spectrum with secondary carriers in the unlicensed spectrum.

The features that can be found in LTE-A technology are:

– *The peak data rate:* LTE-A supports significantly increased instantaneous peak data rates. At a minimum, LTE-A supports enhanced peak data rates to take charge of advanced services and applications (200 Mbps for high and 1 Gbps for low mobility as, shown in Table 4.1).

– *Mobility:* the system supports mobility across the cellular network for various mobile speeds up to 350 km/h (or perhaps even up to 500 km/h, depending on the frequency band). System performance is enhanced for 0–10 km/h and preferably enhanced but no worse than Evolved Units Terrestrial Radio Access (E-UTRA) and Evolved Universal Terrestrial Radio Access Networks (E-UTRAN) for higher speeds.

– *Enhanced multi-antenna transmission techniques:* in LTE-A, the MIMO scheme has been improved in the area of spectrum efficiency, average cell throughput and cell-edge performances. With multipoint transmission/reception, the antennas of multiple cell sites are utilized in such a way that the transmitting/receiving antennas of the serving cell and the neighboring cells can improve quality of the signal received by the UE and reduce co-channel interference from neighboring cells. Peak spectrum efficiency is directly proportional to the number of antennas used.

– *Layered Orthogonal Frequency Division Multiple Access (OFDMA):* OFDMA is used for radio access for LTE-A (Table 4.1).

A technique known as carrier aggregation is used by the layered OFDMA to combine multiple LTE component carriers (from LTE Release 8) on the physical layer to provide the necessary bandwidth. Thus, the layered OFDMA radio access can fulfill significantly higher requirements with respect to system performance and capability parameters compared to the radio access approach used in LTE Release 8. The continuous spectrum allocation concept (used by layered OFDMA for LTE-A) was adopted by the 3GPP Radio Access Working Group, as the approach is backward compatible with LTE Release 8 user equipment (UE) and can be deployed with IP functionality capabilities, low latency and low cost with the existing Radio Access Network (RAN).

Parameter	Antenna configuration		LTE	LTE-A
Capacity (bps/Hz/cell)	DL	2-by-2	1.69	2.4
		4-by-2	1.87	2.6
		4-by-4	2.67	3.7
	UL	1-by-2	0.74	1.2
		2-by-4	–	2.0
Cell edge user throughput	DL	2-by-2	0.05	0.07
		4-by-2	0.06	0.09
		4-by-4	0.08	0.12
	UL	1-by-2	0.024	0.04
		2-by-4	–	0.07

Table 4.1. *LTE and LTE-A capacity comparison*

4.1. HetNet in LTE Advanced

Digital information including all types of data generates heavy traffic that needs to be addressed by telecommunication companies.

LTE and LTE-A systems, popularly known as 4G technology, are considered the technology of choice since they provide cost-effective methods for efficient use of the radio spectrum. With the introduction of the new radio access in LTE which is called Evolved UMTS Terrestrial Radio Access Network (E-UTRAN), LTE has significantly improved end-user QoS in terms of throughput, capacity, and reduced user plane latency. It has enhanced user experience by supporting full mobility. LTE/LTE-A also delivers and supports IP-based traffic with end-to-end QoS. Introducing different types of access networks in LTE is essential to cope with the increasing number of mobile broadband data subscribers and bandwidth-intensive services competing for limited radio resources in 4G networks. Since the sole macrocell cannot meet mobile user have requirements in terms of QoS, LTE-A specifications introduced small cells through the addition of low-power base stations to existing eNBs. These small cells, namely picocells, femtocells and WiFi hotspots, can offload traffic from highly loaded macrocells. Furthermore, LTE-A also introduced the use of relay nodes to expand the area of the macrocells and offer cell-edge users better signal quality and prevent them being penalized by the poor signal transmitted from the eNB.

A generic LTE-A HetNet architecture consists of different types of Radio Access Networks (RANs) or cells: a collocated macrocell, picocells, femtocells and relay nodes, as shown in Figure 4.1. The eNBs are interconnected with each other via the X2-interface while they are connected to the core network (CN) of the operator through the S1-interface. The main difference between each RAN and CN is the split in functionality. The CN architecture is referred to as the Evolved Packet Core (EPC). Together, the LTE-A RAN and the EPC can be referred to as the Evolved Packet System (EPS). EPC supports access to the packet-switched domain only, with no access to the circuit-switched domain, and it consists of several different types of nodes.

While the LTE-A RAN uses a flat architecture with a single type of node, a macrocell is an umbrella for different types of small cells (picocells or femtocells) and can connect farther users through the relay nodes deployed. The RAN is responsible for all radio-related functionality of the overall network including scheduling, radio-resource handling, retransmission protocols, coding and various multi-antenna schemes.

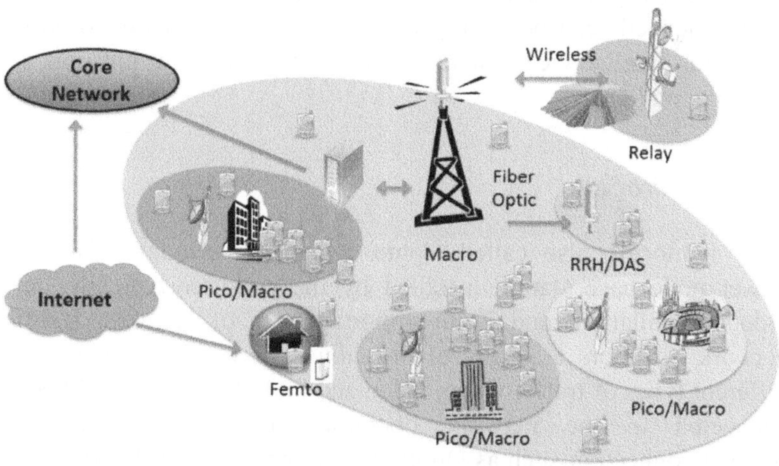

Figure 4.1. *Heterogeneous wireless network architecture*

4.2. Small cell concepts

Different terms have been used by different operators to clarify and distinguish the low-power base stations. Small cell mobile technology is mostly operated in the same licensed spectrum as macrocells. The most frequently nominated unlicensed small cell is WiFi, which provides certain features making it a valuable component of many operators' mobile data strategies.

4.2.1. *Picocell*

Picocell coverage is formed through the use of a low-power node base station called a picocell base station (PBS). It is typically of a smaller order of magnitude to eNB, since PBS integrates an antenna radiating transmission power ranging from 23 to 30 dBm. Omni-directional antennas installed for picocells are different from those used for MBSs sectorial. Moreover, the operator can deploy picocells using the same backhaul and access features as classical eNBs. Picocells can be deployed to fill macrocell gaps or to provide an extra data service where installed, typically in hotspots (for example, malls).

4.2.2. *Femtocells*

A femtocell, also called Home eNodeB (HeNodeB), is a low-power node base station installed by the end-user typically, in an indoor installation, and is an important part of HetNets. It offers network operators a means of providing good indoor coverage for different types of traffic voice, video and high data services. FBSs are connected to the network operator's backhaul via a wired Internet connection such as Digital Subscriber Line (DSL), fiber optic or cable broadband access. The use of femtocells offers several advantages to both the service operator and the end-user. The most important aspect of femtocell design is the access mode mechanism. Femtocells can be configured with three different types of access mode.

4.2.3. *Relays*

Relay nodes represent a low cost solution for extending macrocell coverage and/or improving the throughput of cell-edge users. Unlike other HetNet nodes, such as picocells or femtocells, that are connected to the MBS through X2 or Internet backhaul connections, relay nodes (RNs) use a wireless link to connect to the MBS. Generally, a RN can be considered as a access point that decodes/sends data or forwards/broadcasts information received from a base station in this case it is called Donor eNB (DeNB) or type-II RN, as it does not have a cell ID or a realistic mobility model (RMM). A type-I RN is considered as a small base station with its own cell ID, broadcasts information, and can proceed on RMM based on its own reference signals.

Types	Transmission power	Coverage	Capacity	Connection to backhaul
Macrocell BS	46 dBm	Few km	<1,000 Users	Dedicated wired
Picocell BS	23–30 dBm	< 300 m	10–50 Users	Dedicated wired
Femtocell BS	10–23 dBm	< 30 m	4–8 Users	Wired Internet access
Relay	23–30 dBm	< 300 m	10–50 Users	Wireless

Table 4.2. *Major characteristics of different cells in HetNets*

4.3. Femtocell and macrocell integration architecture

When there is an integrated architecture of femtocell and macrocells, we can consider a geographical area that is totally covered by LTE macro and femtocells. Within the macrocell, an eNB coordinates all data transmissions for UE while in each femtocell the HeNB is the principal entity controlling transmission in the cell. Generally, a UE is able to connect to each network if it is in the corresponding coverage area and perfect power control is assumed to ensure a uniform transmission rate across the coverage area. The architecture used to integrate both macro and femtocells is illustrated in Figure 4.2. The E-UTRAN, which is considered as an access network in the LTE network, permits us to connect the eNB and HeNB to their gateways. The HeNB gateway or the S-GW routes and forwards user data packets between both networks through MME. MME in our case can be considered a centralized entity that controls the resource allocation and interference mitigation between both networks.

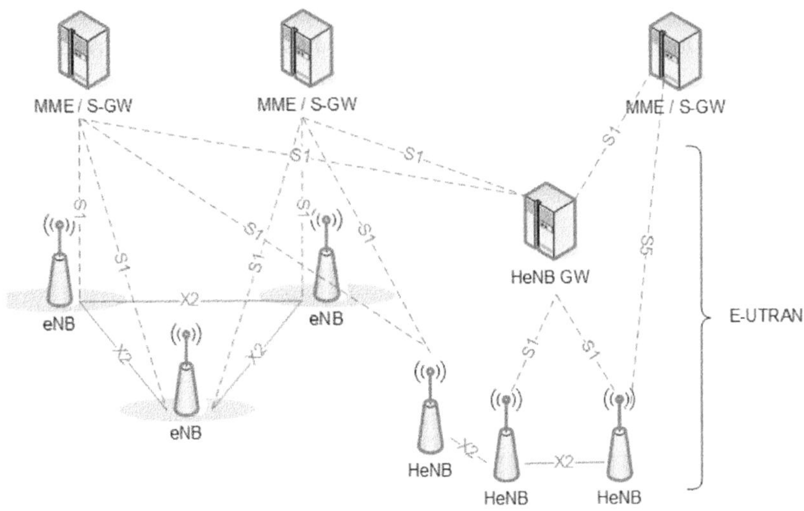

Figure 4.2. *Femtocell and macrocell integration architecture*

4.4. Picocell and macrocell integration architecture

Picocells are one of the key solutions which can be deployed efficiently in saturated areas and can improve the overall system capacity of a region covered by scarce macrocell penetration regardless the coverage type (indoor or outdoor). Despite its benefits, the classical cell selection procedure, which is achieved by comparing and selecting the cell with the maximum received signal strength (RSS) as the user serving cell, introduces uneven traffic load distribution in the network, since most of the UEs connect to the MBS rather than the PBS. Consequently, the macrocell is loaded, whereas the picocells are offloaded. To encourage traffic offloading from the macrocell to its overlaid picocells in a HetNet, cell range expansion has been widely envisioned, enabling more macro users to be offloaded to the less saturated picocells, enhancing the throughput performance of the entire system.

In order to overcome an inadequate picocell data offloading, the concept of cell range expansion has recently been introduced in 3GPP

LTE-A by defining a positive bias value to be added to the PBS RSS during cell selection. In this way, more MUEs are offloaded to picocells even if the RSSs of the selected picocells are below those of the macrocells (see Figure 4.3).

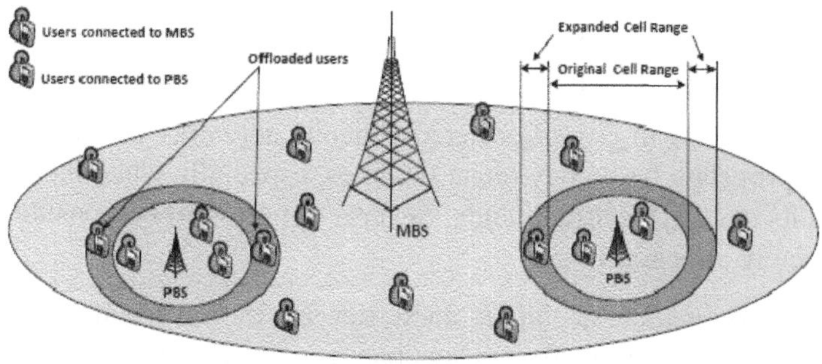

Figure 4.3. *Cell range expansion*

4.5. Interference mitigation in heterogeneous networks

A HetNet consists of a classical macrocell layout with low power nodes placed throughout their coverage zones. Consequently, a HetNet has completely different interference characteristics compared to homogeneous deployment and attention must be paid to these differences when planning frequency reuse. We are interested in the interference generated in two-tier systems, and more especially macrocell/femtocell and macrocell/picocell systems. The interference in two-tier networks is generally of two types:

– co-tier interference, which occurs between cells belonging to the same tier, for example between two neighboring femtocells or macrocells;

– cross-tier interference, which occurs between cells belonging to different tiers, such as femtocell and macrocell or vice-versa.

4.5.1. *Interference mitigation in the context of two-tier macrofemtocells*

Generally, the impact of interference between different tiers, in our case macrofemtocells, depends on the techniques used for spectrum assignment and/or the access method used by the network.

4.5.2. *Frequency spectral assignment*

Interference in cross-tier heterogeneous networks can be mitigated via appropriate frequency reuse schemes, especially when the operator exploits several licensed bands. Several options can be considered for frequency allocation:

1) *Fully dedicated band*: each tier of the overlaid network is assigned a different frequency band; consequently the cross-tier interference is totally avoided. This approach shows low network spectral efficiency because each tier can only access a subset of the frequency resources. If we consider the case of LTE, which is based on OFDMA technology, the terminology "orthogonal allocation" is usually used and the available subchannels are divided into two subsets, one used only by the macrocell tier, the other allocated to the femtocell tier.

2) *Shared spectrum band (universal reuse scheme)*: both tiers share the same frequency band leading to use of total frequency band. Despite this positive aspect (in terms of spectral efficiency), this method produces severe interference that would degrade overall network performance and has to be handled. In LTE OFDMA systems, this allocation is referred to as "co-channel" assignment.

3) *Partially shared band*: is an intermediate solution which provides a tradeoff between spectral efficiency and cross-tier interference by combining the fully dedicated and shared spectrum approaches. In this assignment scheme, macrocell users can access the whole frequency spectrum, while femtocells operate in a subset of it. If some MUEs experience high interference in the shared

spectrum due to their vicinity of the coverage of the other tiers, they are moved to the dedicated band (see Figure 4.4). This approach is referred to as "hybrid assignment" in the LTE system.

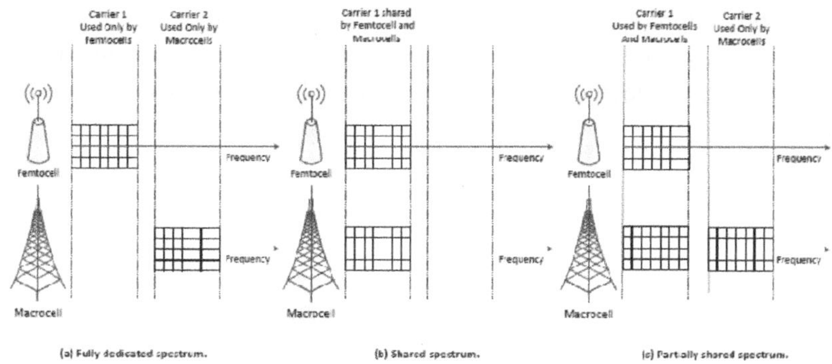

Figure 4.4. *Assignment of carrier frequencies*

4.6. Interference mitigation in the context of two-tier macropicocells

The interference mitigation problem generated in a scenario where the macro and picocells coexist is different from the one presented in the macro and femtocell context. This is mainly due to the open access nature of picocells' deployment as they are managed for the network operator. Unlike the closed access mode femtocells, picocells can always handle macrocell users. Hence, the mobile terminals are always associated with a cell offering a higher quality of Reverse Reservation Protocol (RSRP). Nevertheless, the user association with the highest RSRP might not always be desirable due to:

– *load-balancing*: as the range of picocell areas is much smaller than the macrocells as well as the mobility behavior of most cell phone users, the number of mobile terminals served by the PBSs is much smaller than the number of MBS users. Consequently, the amount of traffic in the picocell may change very dynamically, leading to unbalanced load among different tiers;

– *mobility consideration*: as in any other open-access small cell, cell phone users served by the picocell experience interference from either macrocells and/or neighboring picocells. The mobility behavior causes the mobile user rapidly to change their served cell, or the undesired strength signal is detected where the Radio Link Failure (RLF) procedure is declared and the mobile user desires to reinitiate to the strongest RSRP. Consequently, the number of handover procedures is expected to increase compared to that for homogenous networks.

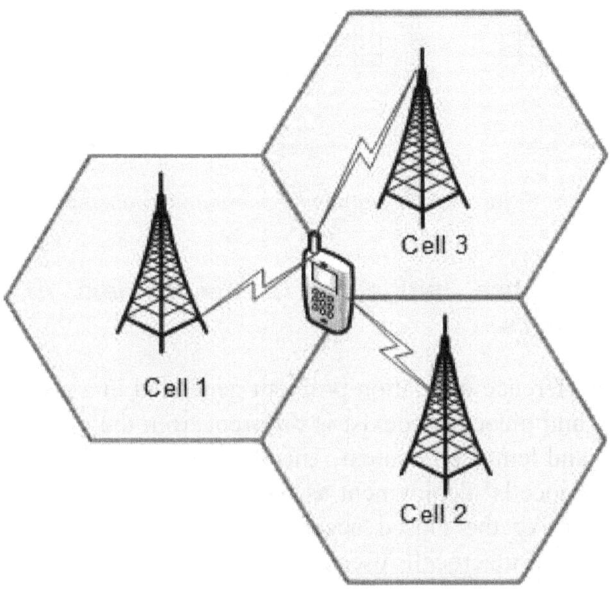

Figure 4.5. *Illustration of coordinated multi-point transmission*

4.7. Coordinated multi-point transmission/reception

LTE-A attempts to enhance the proposed inter-eNB coordination adding coordinated multi-point (CoMP) transmission/reception. Making use of CoMP, several cells may coordinate data transmissions to serve users located between cells who are suffering from low signal qualities. Utilizing this approach, cell-edge intercell interference could be significantly mitigated, because neighboring cells become serving cells. In a similar way, macrocells and femtocells may also coordinate resource assignments utilizing CoMP principles to enhance network performance and mitigate user outages. However, because the same information should be transmitted and/or received at the same time in the same subchannel in neighboring cells, CoMP represents a challenge from the coordination viewpoint. Thus, CoMP schemes involving both macrocells and femtocells are an interesting research topic.

4.8. Carrier aggregation

In order to fulfill the LTE-A requirements in terms of data rate, the 3GPP working group came up with a method of carrier aggregation to support a wider transmission bandwidth for more than 20 MHz which was specified in 3GPP Release 8/9. Carrier aggregation is one of the most distinct features of 4G LTE-A. Carrier aggregation allows the expansion of effective bandwidth delivered to a user terminal through simultaneous use of radio resources across multiple carriers. Multiple component carriers are aggregated to form a larger overall transmission bandwidth. Each aggregated carrier is referred to as a component carrier. The component carrier can have a bandwidth of 1.4, 3, 5, 10, 15, or 20 MHz and a maximum of five component carriers can be aggregated, hence the maximum aggregated bandwidth is 100 MHz. The best way to make use of carrier aggregation is to use contiguous component carriers within the same operating frequency band as specified by LTE, in this case the aggregation is called intra-band contiguous. This might be hard, especially for operators that have their own frequency planning problems or non-contiguous carrier aggregation with some gaps in the frequency carrier see, for example, Figure 4.6.

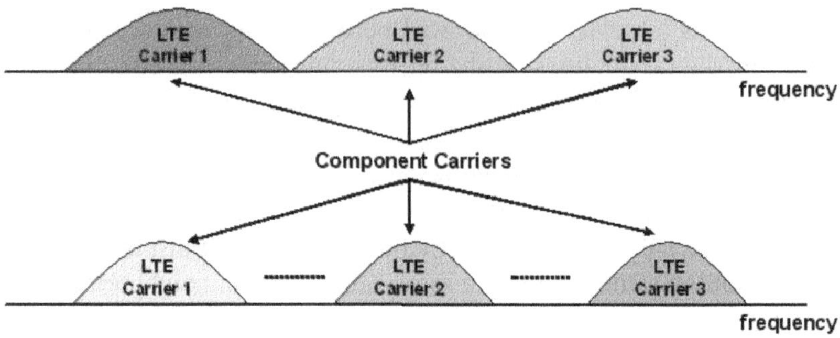

Figure 4.6. *Contiguous versus non-contiguous carrier aggregation*

4.9. LTE Advanced evolution toward 5G

Over the past few years, effort has been made to describe the specification of 5G mobile and wireless technology. No standards gathered all these specifications in one common standard which is why 3GPP started to define the road map for 5G. The last 3GPP release, which is Release 14, started to define what 5G will look like. This initiative is one of the most important and, as stated on the 3GPP website:

> "5G requirements (SMARTER, NexGen) Multimedia Broadcast Supplement for Public Warning System, User Control over spoofed calls, Location services, Mission Critical Video over LTE, MC DATA, LTE support for V2X services, Enhancement for TV Video service, eFMSS, S8 Home Routing Architecture, Ph2 Emergency services over WLAN, Control and User Plane Separation of EPC nodes, Overload Control for Diameter Charging Applications, Latency reduction techniques for LTE, High Power LTE UE for Band 41, Channel model above 6 GHz, SRVCC Enhancements, Service Domain Centralization, Robust Call Setup for VoLTE subscriber in LTE, OAM (energy efficiency and SON for

AAS-based deployments), UICC power optimization for MTC, Requirements for Next Generation Access Technologies, Multi-Carrier Enhancements for UMTS, etc.".

This is why in the following chapter we will cover 5G and try to provide on overview this technology of choice in the near future.

4.10. Bibliography

4.10.1. *Standards*

3GPPTR 36.913 Requirements for Further Advancements for E-UTRA (LTE Advanced), www.3gpp.org.

4.10.2. *Selected bibliography*

[DAM 11] DAMNJANOVIC A., MONTOJO J., YONGBIN W. *et al.*, "A survey on 3GPP heterogeneous networks", *Wireless Communications*, vol. 18, no. 3, pp. 10–21, 2011.

[DEL 10] DE LA ROCHE G., VALCARCE A., LÓPEZ-PÉREZ D. *et al.*, "Access control mechanisms for femtocells", *Communications Magazine,* vol. 48, no. 1, pp. 33–39, 2010.

[GOL 09] GOLAUP A., MONA M., BOONCHIN L., "Femtocell access control strategy in UMTS and LTE", *Communications Magazine,* vol. 47, no. 9, pp. 117–123, 2009.

[HOB 09] HOBBY J.D., HOLGER C., "Deployment options for femtocells and their impact on existing macrocellular networks", *Bell Labs Technical Journal,* vol. 13, no. 4, pp. 145–160, 2009.

[HU 11] HU R.Q., YI Q., SASTRI K. *et al.*, "Hetnets-a new paradigm for increasing cellular capacity and coverage [Guest Editorial]", *Wireless Communications*, vol. 18, no. 3, pp. 8–9, 2011.

[LIN 13] LIN P.-C., RAY-GUANG C., "Dynamic two-threshold flow control scheme for 3GPP LTE-A relay networks", *IEEE 24th International Symposium on Personal Indoor and Mobile Radio Communications (PIMRC),* pp. 2649–2653, 2013.

[LOP 11] LOPEZ-PEREZ D., GUVENC I., DE LA ROCHE G. *et al.*, "Enhanced intercell interference coordination challenges in heterogeneous networks", *Wireless Communications*, vol. 18, no. 3, pp. 22–30, 2011.

[PRA 12] PRAKASH B., NAGATA S., CAMPOY L. *et al.*, "LTE Advanced: an operator perspective", *Communications Magazine*, vol. 50, no. 2, pp. 104–114, 2012.

[REB 14] REBEN K., BOUKHATEM L., MEGUMI K. *et al.*, "Mobility-aware dynamic inter-cell interference coordination in HetNets with cell range expansion", *IEEE 25th International Symposiumion PIMRC,* PP. 1115–1119, 2014.

[REB 15] REBEN K., Cooperation strategies for inter-cell interference mitigation in OFDMA systems, Phd Thesis, University Paris SUD, 2015.

[TAK 09] TAKEDAJ K., NAGATA S., KISHIYAMA Y. *et al.*, "Investigation on optimum radio parameter design in layered OFDMA for LTE Advanced", *IEEE Vehicular Technology Conference*, Barcelona, pp. 1–5, 2009.

4.10.3. *Websites*

http://lteworld.org/blog/carrier-aggregation-LTE Advanced.

http://www.3gpp.org/technologies/keywords-acronyms/101-carrier-aggregation-explained.

5

5G

Along with the efforts of 3GPP to deliver evolved specifications for LTE Advanced (LTA-A), that include many revolutionary characteristics described Chapter 4, there is an initiative within 3GPP itself to define the specifications of 5G in its latest release so this technology will be used in 2020 and beyond. Release 14 is being developed and buzzwords are:

– Internet of Things;

– C-RAN (Cloud RAN);

– full virtualization;

– low bit-cost;

– high spectral and energy efficiency.

These specifications cover new techniques that enable the high data rate to satisfy the high demand of different types of traffic, lower cost bits to be delivered by 5G, and less energy to be consumed to transmit a bit.

Besides these important characteristics, there are many scenarios expected where billions of wireless devices will be deployed and should communicate wirelessly. This will bring a new challenge of a large number of devices with different signaling protocols, diverse types of traffic, different methods of resource allocation, quality of service (QoS), mobility and interference management.

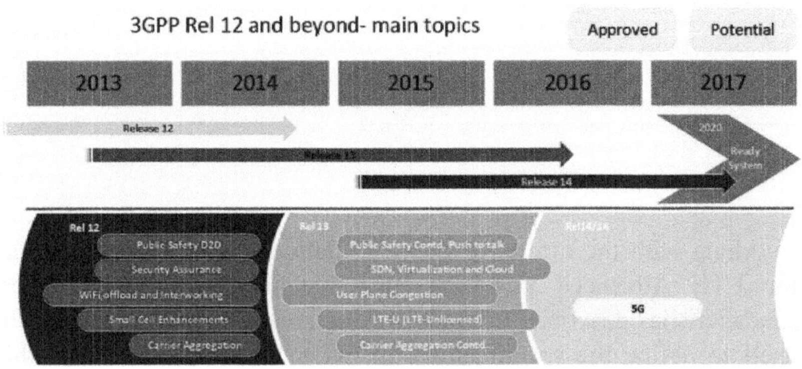

Figure 5.1. *Tentative 3GPP timeline for 5G*

5.1. From LTE Advanced to 5G: the big transition

As shown in Figure 5.1, 5G will be ready in 2020. However, it is difficult to understand what the specification for 5G will be from the figure, and whether there is a gap between the evolution of LTE-A and the first 5G specifications. Work is still in progress on recent releases, especially release 14 and beyond, as LTE-A witnessed the addition of new features that make it the perfect technology of the future. However, it is important to understand the soft transition from LTE-A to 5G as once the releases finish, the next step will be introducing new services on top of features introduced in LTE-A. Thus, one can understand 5G as the enhancement of the radio capabilities used in LTE-A for different types of networks and communication technologies.

Through this chapter, we will show how LTE-A paved the way for 5G. Then, we will introduce the radio capabilities that should be added to evolved elements in LTE-A in order to form what we call 5G.

5.1.1. *D2D communication*

In the LTE Release 12 and beyond, focus will be on network-controlled device-to-device-communication or D2D, where the macrocell base station performs control signaling in terms of synchronization, beacon signal configuration and providing identity and security management. As D2D allows devices in close proximity to communicate directly with each other, they enable authorities to communicate out-of-network-coverage or during network outages (e.g. in the case of a natural disaster). There are also commercial benefits of D2D, with new applications building on the physical proximity of users. The enhanced D2D in LTE-A concerns the introduction of group communications that allow authorities to create and dissolve groups on demand with resource-efficient communications (e.g. multicast) within the group.

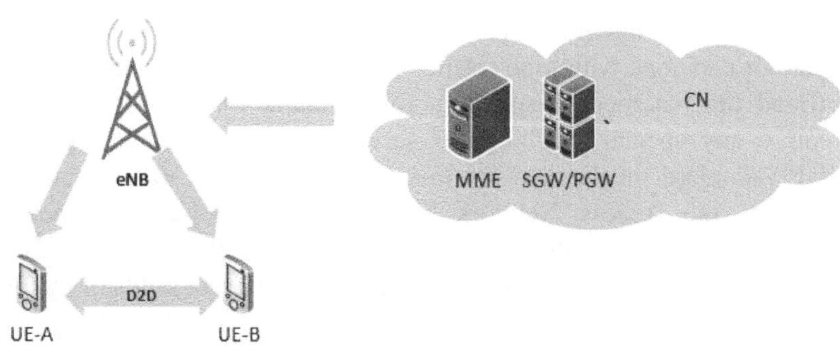

Figure 5.2. *D2D LTE communication for vehicular networks*

5.1.2. *Green activities saving energy*

One of the most important challenges that 5G should address is operator energy consumption switching on their networks throughout the city. This is why one of the characteristics of LTE-A specifications is the use of green concepts. Any operator providing a LTE service uses the energy-saving management functions specified in LTE in order to reduce Tx power, switch off/on cell, etc., based on measurements made in the network showing there is no need. For the full set of network capabilities to remain active. In order to decrease the power consumed, it is important to know how to (1) retrieve energy consumption measurements, (2) retrieve traffic load measurements and (3) adjust network resource capabilities.

5.1.3. *LTE–WiFi integration for traffic offloading*

With the increasing and rapid growth of traffic through the use of billions of smartphones, we need to know how to deal with the huge amount of traffic that an LTE operator has to process. LTE uses small cells, which can be represented by femtocells, metrocells, etc., to offload traffic from macrocells by positioning the small cells either inside the macrocell or at the edge. The best choice is the deployment of WiFi as an integrated part of the cellular network because the operator can cope with high traffic. The 3GPP proposes using WiFi to offload the macrocell, especially for the bulk traffic that does not require any special handling for service delivery or charging. The 3GPP-standard mechanism for this is built around a new functional element, the Access Network Discovery and Selection Function (ANDSF). The ANDSF conveys policies to devices facilitating selection of either cellular or WiFi access for different kinds of traffic, as shown in Figure 5.3.

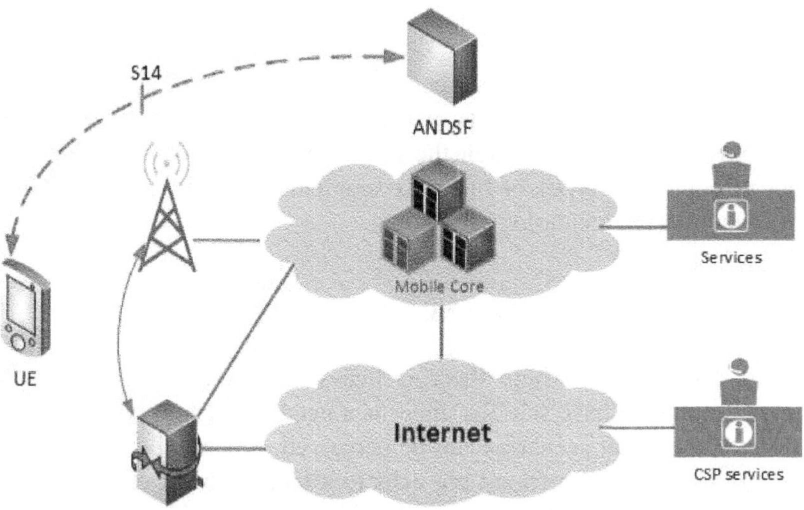

Figure 5.3. *Access network discovery and selection (ANDSF)*

5.1.4. *Vehicular communication*

Vehicular communication (V2X) has many uses, including navigation and driver assistance, travel information, congestion avoidance, fleet management, payment transactions and traffic control and safety. V2X can include many scenarios, just like vehicular-to-vehicular or vehicular-to-pedestrian communication. The use of LTE-A in such a context is advantageous since it provides many capabilities over which many services can be built once it is integrated technically in V2X. For example, V2X moving in all directions is not a big obstacle for LTE-A as it already has extensive coverage in rural and urban area, besides the low latency and high capacity make the deployment of V2X easier. One more important issue is the enabling of D2D communication in LTE that would facilitate the communication of these mobile vehicles by the use of such communications.

5.2. Some characteristics envisioned for 5G

5G mobile and wireless communication system characteristics should meet the requirements of IMT2020. These characteristics will not only be satisfied through the use of new and advanced features, but through new types of business segments that will open once the 5G becomes the technology of choice.

5.2.1. *Massive capacity support*

In dense areas, where there is heavy traffic, many users and a huge number of base stations, there is a need for technologies beyond LTE-A. Different scenarios can be perceived. Small cells can be used as proposed in LTE offload big cells to a large number of small cells according to procedures and algorithms explained in following chapters. However, there is still a problem of license and interference issues that should be addressed. Therefore, 5G attempts to use LTE, which should be suitable for unlicensed spectrum bands where higher frequency bands should be used (up to 10 GHz). As per dense deployment, high frequency bands will be used to create spectrum flexibility by dividing and sharing them among operators in order to optimize the use of the spectrum according to the scenario detected through studying the radio map of the area of deployment.

5.2.2. *Ubiquitous communication support*

One of the most important issues to be covered in 5G is interworking within different types of networks within the same technology, for example making LTE radio more adaptable to various technologies.

5.2.3. *Improvement in radio characteristics*

Regardless the improvement done in LTE-A, there is still more need to improve the capabilities of the radio, especially when we are talking about interworking scenarios for different kinds of networks; i.e. macrocell, femtocell, picocell, D2D, MTC, relay, etc. The coordination of conference in such context is considered to be very hard and new algorithms should be used in 5G to ensure minimum interference. Furthermore, the use of 3D beamforming will be important since it combines horizontal beam pattern adaptation, as applied for beamforming and multiple input multiple output (MIMO) schemes, with vertical antenna pattern adaptation. The recent availability of flexible antenna techniques enables fully dynamic antenna pattern adaptation which can be specified per resource block and piece of user equipment (UE), and makes 3D beamforming practically feasible.

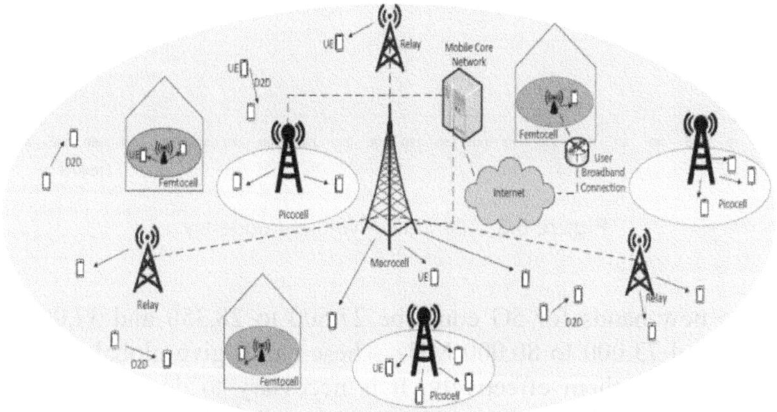

Figure 5.4. *Future vision of 5G*

5.3. 5G frequencies

One of the most important challenges of 5G is to find available frequency bands that allow a large increase in network capacity. One solution is to use cognitive radio, but this is complex and more debates are required to determine the full potential of this technology. The easiest solution is to find frequencies not yet used. Unfortunately, this is almost impossible below 20 GHz so the basic idea is to use very high frequencies in the millimetric bands to provide huge bandwidths. These bands are not easy to use since they have very strong attenuation due to obstacles such as dust or rain. Figure 5.5 shows the attenuations depending on the frequency.

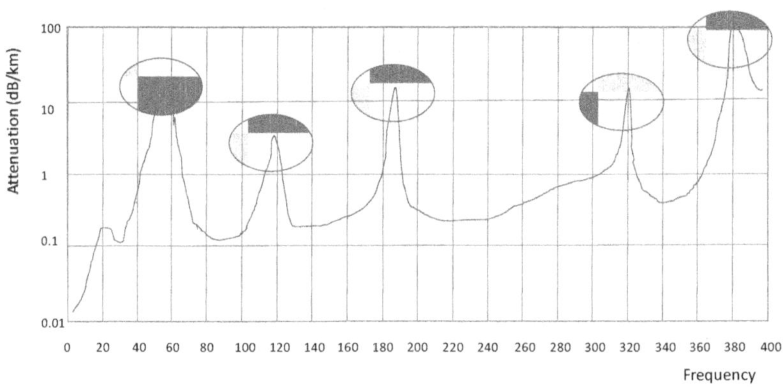

Figure 5.5. *Attenuation versus frequency*

The new bands for 5G could be 27,500 to 28,350 and 37,000 to 40,000 and 73,000 to 80,000 MHz. These bands give global capacity if we can use them effectively. It is necessary to develop accurate beamforming technologies at these frequencies, i.e. get directional signals to reach terminals with high accuracy.

5.4. High and low platforms

5G should also use high altitude platforms (HAP) and low altitude platforms (LAP). The advantage of these platforms comes from a fabulous communication that can cover millions of high-speed terminals. The challenge is to maintain aerial vehicles overhanging the same geographic location. These flying machines can be of different types: aircraft, drone, balloon, etc.

A HAP must be in a stratospheric position (18–22 km). Its coverage is measured in hundreds of square kilometers. The LAP can be a few tens of meters to some kilometers away, the coverage being a few square kilometers. The costs also differ depending on whether one uses inexpensive low-altitude small balloons or a high-altitude sophisticated platform.

As part of the HAP, flying machines are above the flight paths, in an environment almost devoid of turbulance. Balloons move in a circuit, rising and going down, which keeps the balloon roughly stable. A well-known example is Loom from Google, which launched quite a few balloons between 18 and 22 km in altitude in southern Africa and New Zealand. Google's goal is to cover large areas of territory in countries where there is little ground infrastructure. The cover is up to 200 km, at a 5° angle with the ground. Cell partitioning can be achieved when using a large number of antennas. This course allows us to increase the capacity of the platform. Gateways with Earth can be made on frequencies between 5.9 and 7 GHz. Other channels, located in frequencies above 20 GHz, can also be used as for satellite solutions.

These HAPs need to generate their own power to run all network and radio equipment and to propel the balloon. This can be achieved by solar panels, hydrogen fuel cells, etc. Another difficulty for communications is the high power required to cross 20 km or more. Google's platforms, which use wireless frequencies, require a small receiving antenna, which causes additional costs which could be prohibitive in poor areas.

Low-level platforms are much simpler and much less expensive to implement. They may be small balloons or drones flying at an altitude of a few meters to several kilometers. WiFi technology is well adapted to these altitudes, and does not require special antennas to receive the signal.

Mesh networks, detailed in Chapter 13, are well suited to interconnect platforms at similar altitudes, as well as to interconnect at low and HAPs with each other and with geostationary satellites.

A network of drones brings to a large venue or a large event the opportunity to connect to the Internet or offer network access. The drones can provide adequate services to mobile events. Drones can carry enough powerful servers to store, calculate and network equipment. The network of drones forms a flying Cloud that can fulfill many different missions ranging from simple access to complex civil security or military actions.

5.5. Cloud-RAN

C-RAN, or Cloud-RAN or Cloud-Radio Access Network, was introduced in 2010 by China Telecom, with the aim of changing access and backhaul networks between the core network and terminal equipment. This architecture is based on the Cloud: all the control, management and service algorithms are handled by central data centers. This architectural arrangement could have been reached by continuing to use conventional access networks connecting clients to eNodeBs but is fairly revolutionary in the centralized version: the access network is eliminated and replaced with a very simple solution whereby the radio signal is sent directly to the data center.

The radio signal is transferred to the Cloud as it is, without being embedded in a packet. The signal processing takes place in a central data center. The original C-RAN scheme is shown in Figure 5.6. The terminal equipment is connected to the antenna, which retransmits the signal to the central data center. The signal is treated first by a virtual machine executing some signal processing. Then, several other virtual machines execute software associated with the terminal equipment.

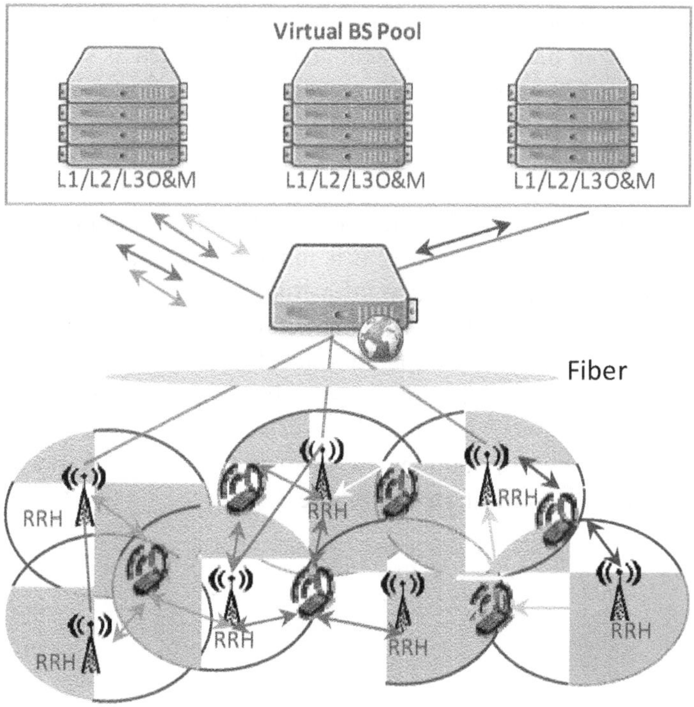

Figure 5.6. *The fully-centralized Cloud-RAN architecture*

When received by the antenna, the signal is sent over an optical fiber, using a technique known as Radio over Fiber (RoF), to the data center. The major advantage of RoF technology is that the same antenna can be used to handle very different signals, such as 3G, 4G, 5G or WiFi. The data center unscrambles the signals and determines their characteristics in order to decode them. Some other advantages are essential, such as the low attenuation of the signal over the fiber. The complexity of the access network is eliminated, and the data center handles this complexity with appropriate virtual machines.

The main disadvantage stems from the use of fiber which is necessary for the communication. WiFi access points have to be connected with optical fiber. This optical fiber is connected to the data center by a BaseBand Unit.

C-RAN uses network function virtualization, i.e. functions decoupled from network equipment and implemented as software by a data center or by a sufficiently powerful server. For a WiFi access point, this means that the WiFi NFV functions are deported to the central data center.

While the C-RAN architecture is interesting for countries wishing to develop a new infrastructure, it is much less attractive to countries which already have a local-loop infrastructure for RAN. In such cases, the proposal is to keep the existing structure and add Cloudlets near to the antennas to perform computation locally. The computation allows local signal processing, equipment signaling and algorithm processing to manage handover, attachment, the choice of connection to be favored when multiple options are available and network operations. This architecture is illustrated in Figure 5.7.

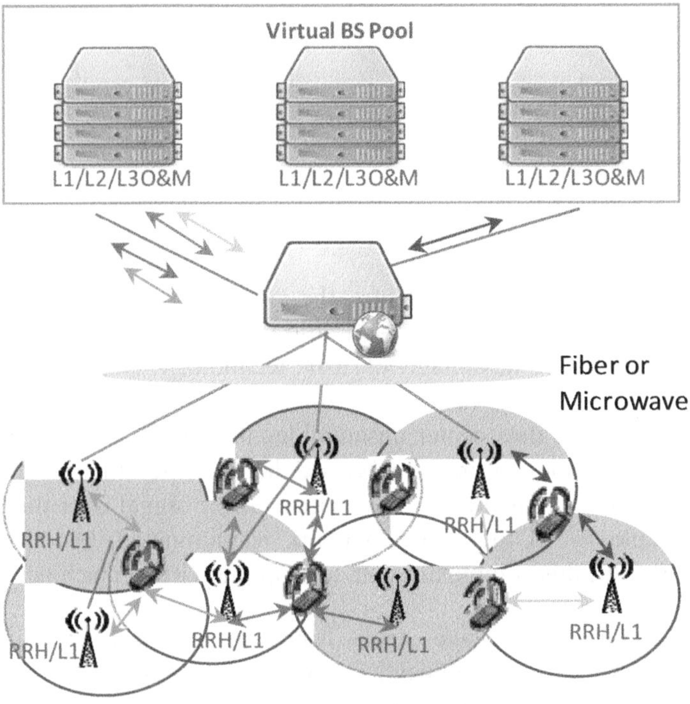

Figure 5.7. *The partially-distributed Cloud-RAN architecture*

5.6. Bibliography

5.6.1. *Standard*

3GP 12 3GPP Green activities/Energy Saving V0.1.0, 2012.

5.6.2. *Selected bibliography*

[AST 13] ASTELY D., DAHLMAN E., FODOR G. *et al.*, "LTE release 12 and beyond", *IEEE Communications Magazine*, vol. 51, no. 7, pp. 154–160, 2013.

[EKR 14] EKRAM H., RASTI M., TABASSUM H. *et al.*, "Evolution towards 5G multi-tier cellular wireless networks: an interference management perspective", *IEEE Wireless Communications,* vol. 21, no. 3, pp. 118–127, 2014.

[ERI 13] ERICSSON, 5G Radio Access, Research and Vision, white paper, 2013.

[HAL 13] HALBAUER H., SAU S., KOPPENBORG J. *et al.*, "3D beamforming: performance improvement for cellular networks", *Bell Labs Technical Journal*, vol. 18 , no. 2 , pp. 37–56, 2013.

[NOK 15] NOKIA NETWORKS, LTE Advanced Evolution in Releases 12–14, New services to pave the way to 5G, white paper, 2015.

5.6.3. *Website*

http://www.3gpp.org/news-events/3gpp-news/1614-sa_5g

6

Small Cells

Long-Term Evolution (LTE) networks promise to change the mobile broadband landscape with peak data rates of over 100 Mbps, high-speed mobility, reduced latency and the ability to support a variety of real-time applications. However, simply providing LTE coverage is not enough to fulfill indoor service requirements. Therefore, operators need to complement macrocell networks with femtocell deployments more tailored to residential and workplace use.

To understand the importance of femtocells for LTE, it is important to analyze mobile customers' behaviors and to determine the nature of this demand and where it occurs. Traditionally, mobile operators' mission is to deliver services to users constantly on the move who use their cell phones mainly for voice services. With the emergence of technologies such as LTE, fixed mobile convergence (FMC) mobile service usages are changing and new trends are appearing leveraging indoor importance. In such a context, high data rates and coverage are the two main ingredients that each operator should offer to remain competitive. However, operators usually fail to provide high quality services to home users and 45% of home and 30% of business subscribers experience poor indoor coverage. With a macrocellular network, it is very difficult for operators to provide high-quality services and cell coverage to indoor users. It is nearly impossible for operators to deploy a huge number of outdoor base stations in densely populated areas to improve indoor coverage. These concerns emphasize the need of femtocells as indoor solutions.

The concept of a femtocell is simple. It is a cheap base station to be deployed in high volume for residential use, connected to the core network via broadband. Femtocells ensure more users receive peak data rates most of the time, especially inside buildings where the vast majority of mobile broadband data is consumed and where the service quality is lower than outside. One of the LTE network bases is Orthogonal Frequency-Division Multiple Access (OFDMA), which is a shared channel radio technology. Hence, LTE femtocells offer better performance and more bandwidth to users as there are fewer users in a cell. Furthermore, LTE femtocells preserve signal degradation and increase throughput since the user device has a smaller cell radius and is closer to the radio.

From a business perspective, femtocells save operational expenditure (OPEX) on the macro backhaul network as they offload traffic from the macrocell network. Capital expenditure (CAPEX) is also saved since no new base stations or capacity expansions are needed. Femtocells also allow operators to maximize LTE spectrum efficiency. New spectrum will be made available for LTE in the high frequency bands that do not penetrate buildings effectively but are ideal for femtocells. This combined with inexpensive voice services will increase revenues.

Femtocells are also designed to offer innovative services that are expected to epitomize LTE. LTE femtocells will provide the best possible environment for downloading and streaming media from the Internet or between devices in the home without loading the mobile network at all. When sharing media in the home, femtocells will not even require broadband backhaul and will therefore not be limited by throughput restrictions on the network, thereby capitalizing on the full peak rates of LTE. Additionally, presence-based applications are envisaged, enabling femtocells to automatically trigger these applications when a consumer is detected entering or leaving the home.

With almost 60% of the worldwide population equipped with cell phones, mobile cellular communication is one of the fastest-growing technologies ever. Recent studies show that voice revenues are declining in favor to data volumes and revenues. This is partly due to the convergence between mobile and Internet since the introduction of third-generation mobile services. With fast and reliable access to the Internet, data volumes have increased far faster than revenues and this trend is expected to accelerate in the future. In order to be competitive, operators need to find ways to substantially decrease the cost per bit of delivering these data, while not placing limits on customers' appetites for consuming the data. Besides voice revenues diminishing, a new trend has appeared regarding wireless usage. Roughly 66% of calls initiated from cellular handsets and 90% of data services occur indoors. Voice networks are engineered to tolerate low signal quality, since the required data rate for voice signals is very low (10 kbps or less) whereas data networks require much higher signal quality in order to provide the higher (in multi-Mbps) data rates. Hence, operators need to improve indoor coverage without additional macrocell deployment. Femtocells are a promising solution with limited cost satisfying both the subscriber, who is happy with the higher data rates and reliability, and operators, who increase revenues with no additional deployment of macrocell networks.

6.1. Femtocell technology

A femtocell also called Femto Access Point (FAP) or Home enhanced NodeB as designated by the 3rd Generation Partnership Project for E-UTRAN is not a new concept. It was first investigated in 1999 by Bell Labs. The original idea was to provide a direct equivalent to the WiFi access point but for mobile cellular networks. The idea had been more widely recognized by 2007 when a number of major companies demonstrated the system at the Mobile World Congress in Barcelona. To promote worldwide femtocell deployment, the Femto Forum was founded in the same year. It comprises more than 100 telecom hardware and software vendors, mobile operators, content providers and start-ups.

From a technical view, a femtocell is a low-power wireless access point, installed by customers themselves in a plug and play way for better indoor voice and data reception. Femtocell operates in the licensed spectrum and is connected to a mobile operator's network using residential DSL or cable broadband connections, as shown in Figure 6.1. Femtocells enables fixed mobile convergence (FMC) service by connecting to the cellular network *via* broadband communications. Femtocells radiates very low power (<10 mW) and can typically support 2–8 simultaneous mobile users.

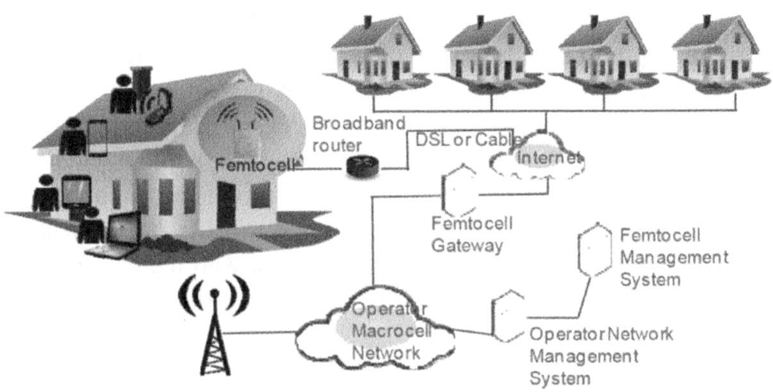

Figure 6.1. *Basic femtocell network*

A femtocell network consists of various supporting network elements: a FAP that provides dedicated coverage to authorized users over the licensed spectrum, and a femtocell access point gateway (FAP-GW) used as a concentrator for all traffic received from the

FAP and the auto configuration server (ACS) used to provide operation administration maintenance and provisioning (OAMP) management functions. The combination of these elements provides communication, security, network provisioning, network management and integration.

The femtocell concept can be applied to different radio access technologies. Although the focus is on LTE, the arguments apply equally to existing or emerging wireless broadband technologies such as 3G and WiMAX:

1) *Femtocell benefits*: femtocells promote a win–win approach where both users and operators could benefit and has attracted strong interest within telecommunication industry.

2) *User benefits*: from the user perspective, femtocells address a number of problems inherent in cellular technology. One of the main advantages is increased indoor coverage with a radius of 50–200 m. In small office/home office environments, femtocells will provide plenty of coverage and resolve the lack of coverage inside buildings. Femtocells improve user quality of experience with a higher data rate which results in better in-building voice and multimedia services. Battery life is also improved because of the low power radiation. In addition, with low deployment cost, femtocells will provide inexpensive voice tariffs. Indeed, using a femtocell to deliver mobile services is much cheaper for the operator, opening up the possibility of charging less for service at home.

3) *Operator benefits*: from the operator perspective, femtocells will enable operators to deal with the increase in mobile usage indoors. Femtocells constitutes increase network capacity by offloading traffic from macrocells. Due to femtocells, the traffic is backhauled to the core network over the existing broadband link without any cost to the operator. The introduction of femtocells has the advantage of increasing average revenue per user and reducing the capital spent per user on new macrocell equipment at the same time. Femtocells enable

emerging data-intensive applications that are a challenge to access from the home via today's cellular networks.

6.2. LTE femtocell architecture

One of the important changes for the evolution in the radio access network (Evolved Universal Terrestrial Radio Access Network E-UTRAN) for LTE is the flat architecture where more functions are added to the base stations, i.e. eNodeB. Currently, the standardization work of femtocells is in progress under 3GPP. Aside from 3GPP, some alliances like the Femto Forum and the Next Generation Mobile Network are rigorously discussing efficient development and deployment opportunities related to LTE femtocells. Figure 6.2 shows the LTE femtocell architecture proposed by 3GPP. The elements of this architecture are listed below.

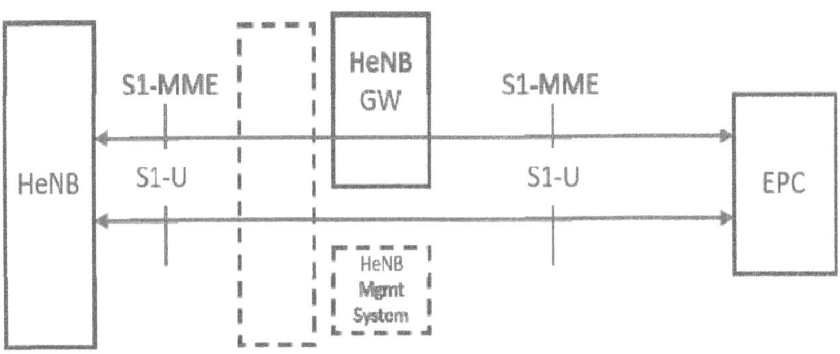

Figure 6.2. *LTE femtocell logical architecture*

6.2.1. *Home eNB or FAP*

Home eNB or HeNB refers to the generic term femtocell in E-UTRAN. HeNB is a plug-and-play consumer device that should be easily installed by the users in home and office environments. The HeNB uses the subscriber's broadband backhaul to connect to the operator's core network. No X2 interface exists between neighboring HeNBs.

Similar to the eNB, the HeNB interconnects with the LTE EPC *via* the S1 interface. More specifically, the HeNB is connected to the mobility management entity (MME) via S1-MME that provides the control functions for idle mode UE reachability and active mode UE handover support. The HeNB is connected to the serving gateway (S-GW) via the S1-UE interface.

6.2.2. *HeNB gateway or FAP-GW*

The HeNB gateway is connected to the HeNB by the S1 interface, and connected to the MME and the S-GW using the same S1 interface. HeNB gateway plays the role of a concentrator and a distributor. The HeNB gateway serves as a concentrator for the control plane, specifically the S1–MME interface at transport network layer. It transports many S1 Application Protocol (S1AP) connections generated by a large number of HeNBs in single Stream Control Transmission Protocol (SCTP) association between HeNB gateway and the MME. SCTP is the protocol used for signaling transport. As a distributor, HeNB gateway distributes messages and traffic to different HeNBs with its range.

The LTE femtocell architecture may deploy a HeNB gateway to allow the S1 interface between the HeNB and the EPC to scale to support a large number of HeNBs. The S1-UE interface from the HeNB may be terminated at the HeNB gateway, or a direct logical U-Plane connection between HeNB and S-GW may be used.

6.2.3. *HeNB management system or ACS*

The functionality of the HeNB Management System (HMS) is based on the TR-069 family of standards. It empowers operators to control and manage the configuration of HeNBs. Furthermore, it produces fault reports and collects different variations in performance from the HeNBs. With the HMS, an operator grants access to HeNBs with additional services and applies service usage policies.

6.2.4. *Security gateway*

In a LTE femtocell, a SeGW is used to provide a secure communication link between the HeNB and the EPC. The SeGW provides protection against potential security threats and network attacks that may occur when mobile traffic is exposed to the public access network. SeGW is a logical entity that comes, in some cases, as an integrated solution within a HeNB gateway, while in other cases it is a separate network entity.

6.3. LTE femtocell deployment scenarios

3GPP, in Release 9 specifications, has proposed three different architecture scenarios for future LTE femtocells.

SCENARIO 6.1.–

Figure 6.3 depicts the first possible architecture according to the 3GPP design. In this architecture, the HeNB only connects to a single HeNB gateway. The HeNB gateway is considered an operator device and is placed in the operator's network. The presence of HeNB gateway enables wide-scale HeNB deployment. This architecture is attractive for vendors, because of its pretty straightforward deployment scenario.

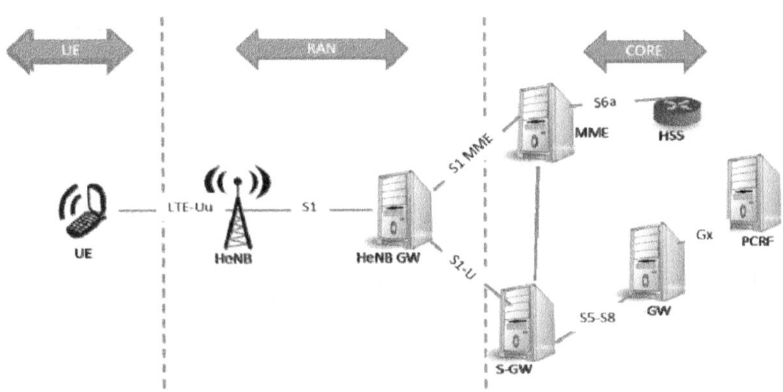

Figure 6.3. *LTE femtocell deployment scenario 1: with a dedicated HeNB gateway*

SCENARIO 6.2.–

Figure 6.4 illustrates another variation of the femtocell architecture without physical existence of the HeNB gateway. In this architecture, the HeNB gateway functionalities are integrated in HeNB and MME.

This architecture follows the principle that in order to increase system performance and efficiency, functional operations of different network devices are integrated in one device in such a manner that the newly developed network entity can perform better under certain conditions with other network entities. This architecture also enables the HeNB to be self-configurable. However, distributing the HeNB gateway functionalities in the HeNB and CN might cause performance degradation in HeNBs.

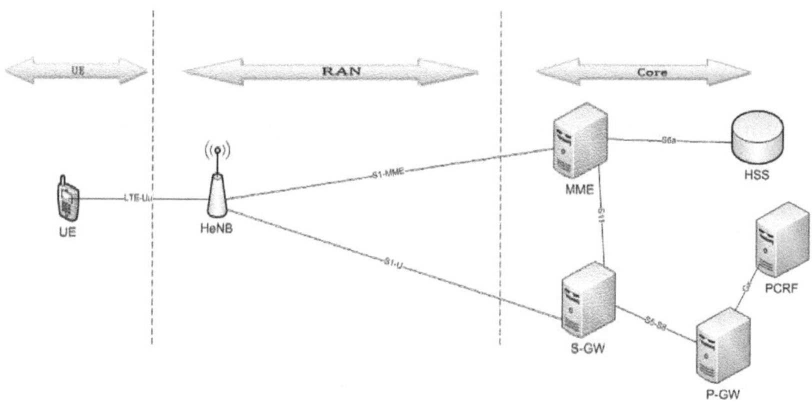

Figure 6.4. *LTE femtocell deployment scenario 2: without a HeNB gateway*

SCENARIO 6.3.–

In this architecture variation, the HeNB gateway is only used to aggregate control plane signaling and the HeNB is directly connected to the S-GW using a direct logical U-Plane connection, as illustrated in Figure 6.5. In this way, the HeNB gateway is used for transporting control plane signaling, efficiency of data packet delivery to the S-GW will increase and this will increase the overall data packet transport efficiency in the whole network.

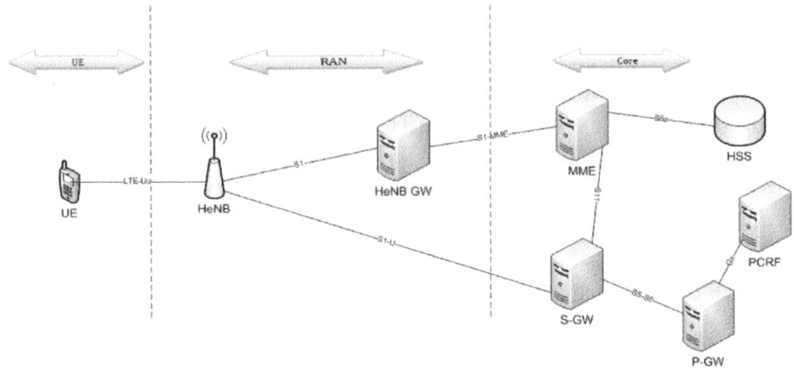

Figure 6.5. *LTE femtocell deployment scenario 3: HeNB gateway for contro plane*

6.4. Femtocell access control strategy

Access control is one of the key ways of supporting various features of the femtocell, including mobility support, interference management and so on. The ability of UE to distinguish between macrocell and femtocell is of a great importance as it limits their battery consumption.

6.4.1. *Closed subscriber group*

3GPP, in Release 8, introduces the concept of the closed subscriber group (CSG) which consists of identifying a group of subscribers who are permitted to access one or more Public Land Mobile Network (PLMN) cells. This restriction is imperative for femtocells for several reasons; femtocells usually support a small number of user devices as more devices will provide inadequate service. Furthermore, the femtocell owner would not agree to share its backhaul link with other users.

The CSG concept enables permitted UE to use the femtocell by defining a CSG identity that is unique in the PLMN. The CSG identity is broadcasted by all femtocells that support authorized UE access.

Each piece of UE should be able to store a list of allowed CSG identities.

6.4.2. *Femtocell access control modes*

Femtocells can be used in three different usage models:

– *Open access mode*: In the open access method, all UE is treated equally from a camping and charging perspective. A provider deploys a femtocell to provide coverage in an area where there is a coverage hole. In this instance, it would be beneficial to allow public access to the femtocell; this is a hotspot scenario like that in a coffee shop or airport. In this model, the femtocells become another part of the PLMN. The disadvantage of this method is that it may increase the number of handovers and signaling and it may also face some security issues.

– *Closed access mode*: in closed access mode, only CSG subscribers are allowed to connect to a privately accessible femtocell. UE that is not part of the CSG would not get access to the femtocell except for emergency calls. A closed access mode is also referred to as a CSG cell in 3GPP terminology.

– *Hybrid access mode*: the hybrid access approach, also known as a hybrid cell in 3GPP terminology, is similar to the closed access mode. It allows UE not part of the CSG to connect to the femtocells, but only a certain portion of the resources with preferential charging and access for the subscribers. UE not part of the CSG may be pre-empted or rejected in favor of subscribers.

6.4.3. *Physical cell identity*

In 3GPP, each LTE cell broadcasts a specific physical cell identity (PCI) that is normally used to identify a cell for radio purposes; for example, camping/handover procedures are simplified by explicitly providing the list of PCIs that UE has to monitor. This list is usually known as the neighboring cell list. The PCI of a cell does not need to be unique across the entire network; however, it must be unique on a local scale to avoid confusion with neighboring cells. In the 3GPP radio access technologies, a range that includes PCIs for exclusive use by femtocells can be signaled. This represents a challenge in femtocell networks, since they must select their PCIs dynamically after booting or changing their position in order to avoid collision with other macro/femtocells. Furthermore, in extensive femtocell deployments and due to the limited number of PCIs, 504 in LTE, the reuse of PCIs among femtocells in a given area may be unavoidable, thus causing PCI confusion. A subset of the PCIs can be reserved for specific purpose, such as identification of the femtocells. For instance, a PCI can be used to identify a CSG cell without reading the system information by attributing a specific range of PCIs to the CSG cells. These sets of PCIs will be exclusively used by CSG cells and UE not interested in accessing the CSG and can avoid camping on them while preserving battery life. However, if a fixed number of PCIs are reserved for a specific use, PCI confusion be severe when the large number of femtocells is deployed. A dynamic PCI reservation scheme could be a good solution to this issue.

6.5. LTE femtocell challenges and technical issues

The femtocell solutions are overwhelmingly attractive to mobile network operators as they successfully address coverage and mobile data bandwidth requirements by leveraging widely-available broadband connections without the extraordinary cost associated with the alternative macrocell deployment. Even though usage of femtocells provides many benefits, it poses serious challenges to LTE femtocell deployment due to several technical issues.

6.5.1. *Interference*

A major challenge with femtocell technology is the interference that can be caused by a large deployment of HeNBs associated with the existing macrocell. Femtocell deployment can cause severe degradation of the neighboring cells if not managed properly. Since femtocells and macrocells are located in two different layers, the resulting network is known as a two-tier network. In two-tier networks, we distinguish two types of interferences: cross- and co-tier interferences. Cross-tier interference is caused by an element of the femtocell tier to the macrocell tier and vice versa. Co-tier interference takes place between elements of the same tier, such as neighboring femtocells.

Some techniques have been proposed to overcome interference in femtocells. We can distinguish hardware-based approaches, such as cancelation techniques, or the use of sectorial antennas. However, these techniques usually imply an increase in HeNB cost which is contrary to the essence of femtocells.

Efficient alternatives are represented by strategies based on interference avoidance and subchannel management. These techniques are often used to mitigate interference in cellular networks and are of great importance in femtocells due to cross-tier interference.

6.5.2. *Spectrum allocation*

From the point of view of OFDMA spectral resource allocation, different approaches exist to manage subchannel allocation. One way to completely eliminate cross-layer interference is to divide the licensed spectrum into two parts; this is called orthogonal channel assignment. This way, a fraction of the subchannels would be used by the macrocell layer, while another fraction would be used by the femtocells. In orthogonal channel assignment, the spectrum allocation can be static, depending on the geographic area, or can be dynamic depending on traffic demand and user mobility. Although optimal from a cross-layer interference standpoint, this approach is inefficient in terms of spectrum reuse and implies that UE should support different frequencies.

Another subchannel assignment consists of sharing the spectrum across the femtocell and macrocell layers; this is called co-channel assignment and is more efficient and profitable for operators due to the high cost of the licensed spectrum. It enables high-frequency efficiency and easy handover between eNodeB and HeNB. However, transmissions within femtocells may cause macrocell cross-interference and vice versa. Resource allocation under co-channel deployment is thus of critical importance. The key to the success of such a strategy assignment is to protect macrocell user services against femtocell interference while exploiting spatial channel reuse.

Two strategies can be employed to mitigate cross- and co-tier interference. In a centralized strategy, there would be a central entity that is in charge of intelligently assigning each cell subchannel. This entity would collect information from the femtocells and their user and use it to find an optimal solution within a short period. However, with salient features of femtocells such as user installation and unplanned deployment, this solution is inappropriate due to hard technical barriers.

A distributed and self-organizing strategy is also being investigated to mitigate cross- and co-tier interference, where each cell manages its own subchannels in a cooperative or non-cooperation manner. In a non-cooperative approach, each femtocell would plan its subchannel in order to maximize the throughput and QoS for its users independently from the effects its allocation might have on neighboring cells. Access to the subchannels then becomes opportunistic, and the method decays to greedy. In a cooperative approach, each HeNB gathers information about neighboring femtocells and may perform its allocation taking into account the effect it would have on neighbors. In this way, the average femtocell throughput and QoS, as well as their global performance, can be locally optimized.This approach would be very beneficial, both in terms of resource management and interference reduction. But, it has the disadvantage of requiring additional overhead or sensing mechanisms to gather information about neighboring femtocells and macrocells.

6.5.3. *Access mode impact*

Another factor that plays a key role in the severity of interference is the access strategy within femtocells. In a closed access scenario, non-subscribers can experience severe jamming from the nearby femtocells. Non-subscribers are not allowed to use the femtocell even if the femtocell pilot power is higher than that of the nearest macrocell. Nonsubscribers can also jam the HeNB uplink if they are transmitting with high power. In an open access deployment, non-subscribers can also connect to femtocells, which considerably reduce cross-tier interference. However, since all users can use femtocells, the volume of signaling messages increases due to the high number of handovers. Open access can thus create outages.

6.6. Security and privacy challenges

Like all communications technologies, femtocells require robust security. The threat model for a femtocell-enabled mobile network, as illustrated in Figure 6.6, consists mainly of attacks on the communication links between UE, HeNB and carrier's core network and attacks on the integrity of the HeNB, which is considered a vulnerable point in a femtocell network.

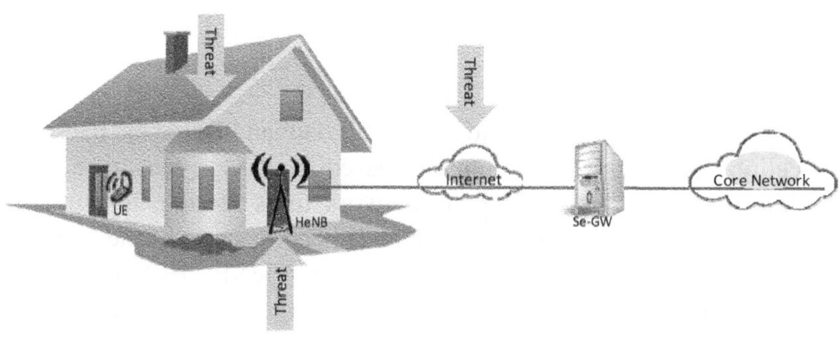

Figure 6.6. *Threat model for femtocell networks*

1) *Attacks on communication links*: attacks on telecommunication links or third-party attacks on networks intend to compromise the security between UE and HeNB and/or the HeNB and the SeGW for the purpose to of eavesdropping, service disruption, fraud and other malicious activity. This kind of attack includes man-in-the-middle, traffic snooping and tracking attack.

The communication links in a femtocell network can be divided into:

– the link between the UE and the HeNB through the air interface;

– the public link between the HeNB and the SeGW.

Attacks on the air interface can be either passive or active. In passive attacks, the attacker passively listens to communication whereas in active ones, the attacker injects or modifies the data in addition to listening. Attacks on the air interface are being tackled by cryptographically protecting the messages sent over the air.

The issue of user identity protection is raised in migration toward all-IP and femtocell networks where the legacy solutions (TMSIs and GUTIs) might not be suitable. With the massive deployment of femtocells, the use of unlikable temporary identifiers to protect the identity of mobile devices at the air interface is insufficient to guarantee a satisfactory level of protection since they are usually unchanged in a location area composed of up to 100 adjacent cells. Femtocell thus enables tracking with unprecedented accuracy due to the low range.

Since HeNB and HeNB gateway use broadband IP for the backhaul of cellular communications, Internet-based attacks are feasible against mobile operators. HeNB and the network must be able to mutually authenticate each other so that the HeNB becomes part of the carrier's network. To enable this process to occur, HeNB, HeNB gateway and SeGW that sit between the public Internet and the mobile operator's core network must be able to establish a secure communication tunnel.

Standardized by the Internet Engineering Task Force (IETF), IKEv2 has been prescribed for the HeNB and HeNB gateway authentication requirements of addresses. It is a flexible protocol that supports many actual authentication methods. Authentication within IKEv2 can be performed with Public Key Infrastructure (PKI) certificates, shared keys or even SIM cards.

IKEv2 also supports the Extensible Authentication Protocol (EAP), which is a key feature in applying the IKEv2 protocol in many existing authentication schemes or systems. After successful negotiation, identification and authentication of all parties, IKEv2 generates the keys and establishes connection for further secure communication.

While IKEv2 is used to authenticate the access points and gateways for each party, the actual secure communication channel is realized with IPSec. This is another IETF standardized protocol for securing Internet communications. Support for the IPSec protocol is a requirement for protecting the IP backhaul of the femtocell system. IPSec protects IP traffic as it travels over the broadband connection back to the carrier's core network. It is a flexible and efficient method of providing data integrity and confidentiality. While IPsec is a complex suite of many protocols, backhaul security within femtocell networks focuses specifically on one variant, the Encapsulating Tunnel Payload (ESP) tunnel variant.

2) *Attacks on the integrity of the HeNB*: HeNB is a gateway to the carrier core IP network and the carrier's radio network. The most common threats to HeNB integrity include hacking, tampering/reverse engineering and device cloning. HeNBs are prone to attacks regarding the physical size, material quality, lower cost components and the IP interface. In addition, the standards for femtocell call for radio configuration data, encryption keys, identity material and operational

statistics to be stored within the HeNB itself. These data are sensitive and must not be available to any party but the carrier. To achieve this, the data must be stored in a robustly protected "cryptographic safe" within the device. When the HeNB is physically vulnerable to tampering by malicious users, attacks such as device impersonation, IP attacks on the network services and false location reports have severe consequences on quality of service. The validity of billing, subscription and device data must be secured. To this end, HeNB should be equipped with a trusted execution environment.

6.7. Synchronization

In 3GPP specifications, base stations use frequencies that are very accurate and closely synchronized. Network time synchronization is particularly important between macrocells and femtocells in order to minimize multi-access interference, as well as for the proper performance of handovers. This issue is critical since femtocells are deployed by users and there is no centralized management of their radio resources. Without timing, transmission instants would vary between different cells. This could lead to the uplink period of some cells overlapping with the downlink of others, thus increasing intercell interference in the network.

For a low-cost femtocell implementation, synchronization is becoming significant. The manufacture of low-cost femtocells equipped with high precision oscillators is not trivial, so alternative approaches need to be considered in order to achieve reliable time synchronization. One solution is the use of the IEEE-1588 Precision Timing Protocol shared by standard for an accurate clock synchronization protocol between network equipment as a feasible method to achieve synchronization. However, some modifications are necessary for it to perform efficiently over asymmetric backhaul links such as ADSL.

An alternative is the use of GPS receivers, which provide accurate timing over satellite links. Since accurate location information is not required, only a single satellite is needed. Performance depends on the availability of GPS coverage inside user premises.

A third possibility is that the base station could receive transmissions from the overlaying (highly accurate) macrocellular network and adjust its timing accordingly (network listening), although this adds the overhead of another radio and requires a macrocell be within range of the femtocell.

Finally, there are some innovative, low-cost/high-stability Temperature Compensated Crystal Oscillator (TCXO) products coming on the market which might reach the required stability levels and make the issue easier.

6.8. Mobility

In a LTE macrocell, UE mobility support is classified into two states: idle mode and connected mode. In the idle model, cell selection and reselection are done for the mobility management of UE. When UE is turned on, PLMN is selected. The UE searches for a suitable cell within the selected PLMN and chooses the cell to provide available services and tunes to its control channel. This choice is referred as “camping on the cell”. If the UE finds a more suitable cell, according to the cell reselection criteria, it selects onto that cell again and camps on; this mechanism is defined as cell reselection. When a call is generated, the idle mode is transited to the connected mode.

The LTE utilizes a network-controlled and UE-assisted handover procedure for mobility in connected mode. For the handover between two eNodeBs, the handover decisions are made by eNodeBs without consulting the MME: the source eNodeB decides to move the UE to the target eNodeB based on the UE measurement report; the target eNodeB prepares radio resources before accepting the handover; after a successful handover, the target eNodeB indicates the source eNodeB to release the resources and sends a path switch message to the MME. The MME then sends a User Plane Update Request to the S-GW routing information update. The control messages are exchanged via the X2 interface between the two eNodeBs. The downlink packet data are also forwarded from the source to the target eNodeB via the same X2 interface. However, in the femtocell architecture, no X2 interface exists between neighboring HeNBs. Moreover, only authorized users can access HeNB, while in LTE macrocell handovers are performed without restrictions.

Considering deployment of the HeNB in 3GPP LTE systems, there are two types of handover: inbound handover from macrocell to HeNB and outbound handover from HeNB to macrocell:

– *Inbound handover for HeNB*: this is the frequently facing handover scenario for UE and is complex. With the deployment of thousands of possible target HeNBs, inbound handover for HeNB is the most challenging issue for LTE femtocell networks. If UE belongs to a subscriber group (CSG) cell, only UEs belonging to the CSG is permitted to access and receive service from the CSG femtocell. The authentication is thus checked in preparation for handover, which results in more complex handovers than in macrocells.

– *Outbound handover for HeNB*: in this scenario, UE is currently connected with HeNB and it is important which entity, i.e. eNodeB and HeNB, decides the handover from HeNB to the target cell because it affects the handover procedure. The outbound handover is less challenging than the inbound handover because whenever a user moves out of a femtocell network, eNodeB signal strength may be stronger than HeNBs. The selection is easier, as there is no complex authorization check and interference calculation compared to inbound handover. Considering both handover scenarios, the main problems are:

- a method for identifying HeNB is needed regarding the large number of potential femtocells that could be deployed in a macrocell, the accessibility of UEs especially where CSG deployment is also of great importance. If the UE measures all the HeNBs and reports the measurement to the eNodeB, it causes high UE power consumption and increases handover delay. The quantity of measurements for a non-allowed CSG cell should also be avoided;

- management of the neighboring cell list that represents the group of candidate target cells for handover. In LTE system, the X2 interface is defined and directly connected with the eNodeBs. Hence, managing the neighboring relation is simplified.

However, HeNBs and eNodeBs are not directly connected and HeNB is frequently turned on/off. Therefore, managing the neighboring list which consists of both HeNBs and eNodeBs is complex.

Finally, cross-interference between the HeNB and eNodeB is also a key issue that affects handover support in both handover scenarios.

6.9. Passpoint

Passpoint is an architecture proposed by the WiFi Alliance, which was created with the arrival of WiFi to promote this technology. The WiFi Alliance has greatly contributed to the success of WiFi and the introduction of various standards such as WPA2 which allows us to offer a high security in WiFi networks.

Passpoint is a solution that allows us to replace 3G/4G antennas with an offloading solution allowing users to connect in a completely transparent way through WiFi networks. In other words, the customer does not know how it is connected to the operator's network: a BTS antenna, a NodeB or an E-NodeB or by a wireless access point (hotspot, femtocell, picocell, microcell, etc.).

In home or office environments, connecting to a WiFi network is usually automatic once the user has supplied credentials. When connected to the access point and allowed to enter the network, the user is subject to the rules established by the IT manager or the owner of the access point. Once it is recognized, the terminal is automatically associated with the access points to which it connects regularly and without user intervention.

The connection to most hotspots is often different. In a public place where there are several networks, customers must often start by choosing which network they want to connect to and be recognized by the access point. In most cases, the customer enters authentication information. There are solutions to simplify network selection and automatically make the association while ensuring good security, but these are often limited to a small number of networks and are rarely available abroad. WiFi Passpoint technology fundamentally changes the use of public hotspot. This is a general solution that all hotspot operators can apply that overcomes the limitations of proprietary and interoperable solutions offered by existing providers. A program in certified equipment automatically manages the association's network, authentication, authorization and security in a totally transparent way. This solution works in all Passpoint networks.

First, Passpoint handles the discovery and network selection. The terminal automatically discovers and chooses the network on which to connect from a list of preferences specified by users, operator policies and network availability. These characteristics are based on the IEEE 802.11u standard.

Access to the network is done seamlessly, without the user having to do anything. It is no longer necessary that the terminal is on a list of active access points to which it agrees to connect. There is no need to enter account information in a browser. Passpoint uses a consistent interface, and the connection process is automatic. Moreover, the choice of device can be made automatically for multiple types of credentials. Passpoint supports SIM card authentication. These cards are widely used in current cellular networks. Passpoint also supports combinations of user-name password and certificates. No user intervention is required to connect to a trusted network.

All Passpoint connections are secured by WPA2-Enterprise for authentication and connectivity, providing a level of safety comparable to that of cellular networks. Passpoint improves WPA2-Enterprise by adding features to contain known attacks in on public WiFi networks. Passpoint also provides instant open accounts when the user does not have a subscription to a telecommunications operator. This process is standardized and unified to establish a new account at the access point, using the same account opening method used by all WiFi Passpoint ISPs. Passpoint creates a global platform centered on four protocols based on the standard Extensible Authentication Protocol (EAP). This protocol is widely deployed today. The EAP-SIM, EAP-AKA and EAP-TLS authentication methods allow mobile operators to use the same authentication solutions for cellular and WiFi.

In addition to ease of use through seamless authentication, Passpoint brings many benefits to service providers and users:

– Internet access for electronic devices without a browser. The Passpoint authentication method requires no browser, facilitating the use of electronic devices such as cameras, embedded devices in cars, connected objects and more generally all the "Things" that connect to Internet. The Internet of Things is described in more detail in Chapter 11.

– Simple connection and opening of new subscriptions. Whether to connect the subscriber to an existing account or to introduce new services, automation of authentication and security of the connection process make Passpoint an attractive solution for content providers and oriented terminal manufacturer content such as digital tablets. WiFi users can use prepaid subscriptions according to a model similar to that of cell phone operators.

Service providers increasingly want to protect their premium content subscribers. They need to find out who receives such content. Passpoint provides the identity of service providers, subscribers access rights and offers premium quality content for subscribers connected at home, in a corporate network or via public access points.

Passpoint hotspots offer service providers seamless roaming in other operator networks. To activate the roaming service, operators must first establish mutual roaming agreements that cover the validation of access and billing. Roaming is based on a single network selection protocol and user authentication in the hotspot. Once roaming between service providers activated, Passpoint devices can connect automatically to the network using the same process. In all cases, the customer recognizes the Passpoint access point as belonging to the list of available networks and automatically establishes a connection, such as for cellular roaming. The new features are enabled only if the access point and the client device are compatible with Passpoint. The WiFi Alliance Passpoint certification program will ensure compatibility. In addition, Passpoint supports connectivity with non-Passpoint equipment.

Customers with the correct credentials and using a suitable EAP method can connect to Passpoint access points. Customers who do not have proper credentials for EAP methods must use an open system based on browser authentication, where WPA2 may not be necessary. In such cases, security is not guaranteed.

Passpoint clients can connect to any existing WiFi network. However, they will not be able to use Passpoint features and services. For example, standard hotspots offer no connection security, while Passpoint automatically establishes WPA2 encryption.

WiFi Certified Passpoint is a WiFi Alliance certification program for both access points and client devices. It ensures, just as for WiFi equipment, the interoperability of equipment and terminals.

6.10. The backhaul network

Backhaul networks are intermediate networks between access networks, to which customers are connected, and the core network. These backhaul networks are becoming even more important as small cell technology requires a very large number of connections. Backhaul networks essentially make connections between the core and digital subscriber line access multiplexer networks or between base stations and the core network. These networks are generally using Gigabit Ethernetor MPLS solutions.

Mesh networks, where access points are directly connected to each other, need two wireless cards, one for communicating with neighboring access points, the other to connect customers. A client that wants to send packets connects to an access point, which forwards the packets to neighboring access points, which in turn transmits to its neighbors, and so on until it reaches the core network via an optical fiber or an xDSL line. The access points are typically connected by WiFi, but other solutions are being developed with longer ranges, as we will see in Chapter 8.

Figure 6.7 shows a typical configuration of a backhaul network. The small cells are connected to a "Small Cell Backhaul Aggregation" station which recovers the flow rates of different cells and aggregates to convey this flow to the core network, either by optical fiber or by radio relay. The waves are directed to switches in the Radio Access Network (RAN), which are themselves connected to a switch in the core network. This last link is often provided by a broadband multiprotocol local switching solution.

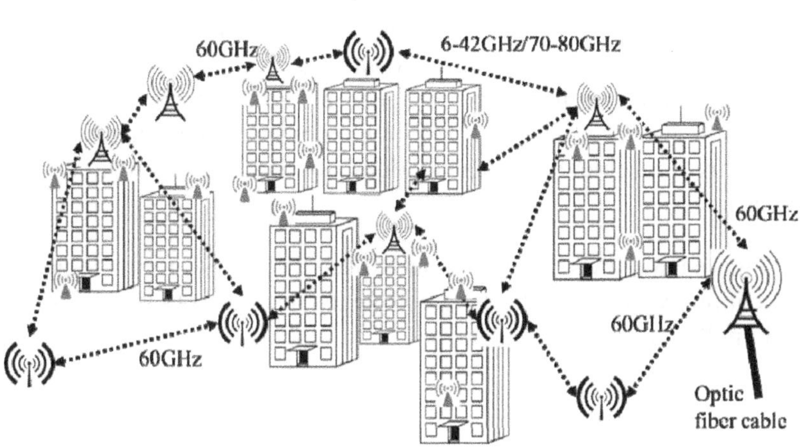

Figure 6.7. *A backhaul network*

The connections between the access points and aggregators, even with the first RAN switch, can be made using a radio link at very high capacity of several gigabits per second on high frequency bands, such as 60 GHz. The major drawback of these frequencies is the need for direct view and favorable weather. Indeed, these frequencies located below the millimeter are blocked by water droplets and airborne dust. Specifically, the connection may be 105 GHz or 70–80 GHz or 57–64 GHz or between 6 and 56 GHz. The privileged band appears to be 60 GHz for various reasons: the available bandwidth, flow rates of a gigabit per second, low interference (because the links are very directional), low latency, small components and antennas and finally low production cost. The disadvantage of this band is the distance is limited to few kilometers between the two antennas to avoid excessive losses.

6.11. Software radio and cognitive radio

Cognitive radio is one of the most important revolutions of the late 20th Century. It uses the radio frequency spectrum well, however optimum use cannot be made without a certain number of advances that we will explains.

Software radio, or software defined radio (SDR), is a transmitter or radio receiver made using a software solution. One immediately sees the implications of such technology since the characteristics of the interface can be changed infinitely without any technical problem. The user can have the interface he needs at a time t and completely different interface at time t + 1. Obviously, some hardware is necessary to achieve the calculations for coding, modulating and filtering.

Figure 6.8 shows what we can expect from these technologies in the coming years. The objects are arranged according to the flexibility and intelligence.

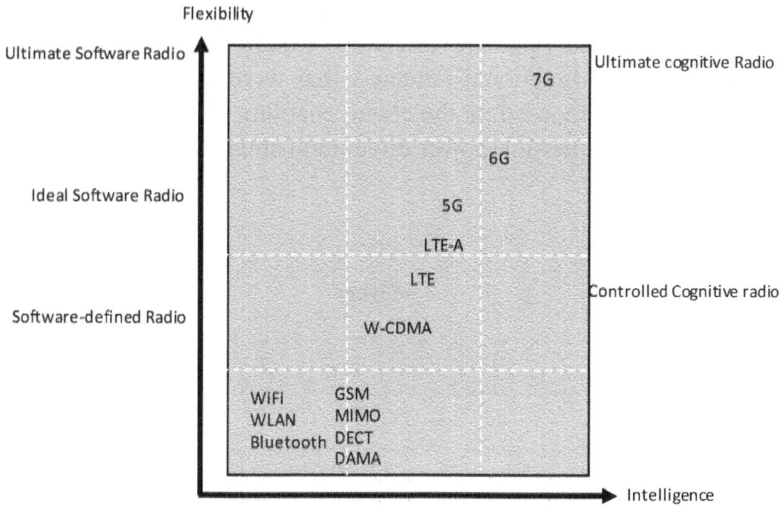

Figure 6.8. *Software radio and cognitive radio*

Using these technologies, we should increase the rates by a factor over 10 years.

6.12. Custom cells

Among the many directions of research and development relating to 5G, custom cell is one of the most promising, allowing a high capacity while reducing at the maximum the power necessary for communication between the user and the station. This solution provides an excellent reuse spectrum. The idea seems simple: the cell in which the user is located must not interfere with another terminal nearby. This solution is illustrated in Figure 6.9.

When the terminal wants to send a packet, the antenna is capable of transmitting directionally from a large number of virtual antennas. We could have a general antenna formed from hundreds of virtual directional antennas. This antenna recovers, as shown in Figure 6.9, the five most powerful waves: direct stream, two streams with one rebound and two additional waves with two rebounds. Once this learning process is completed, the main antenna is able to transmit to the terminal in the opposite direction, exactly repeating incoming waves and using directional antennas that were used to receive the signal. The cell is focused on the client, enabling high-frequency reuse and very low transmission power directional antennas.

Figure 6.9. *A cell on demand*

This solution is somewhat complicated when the user terminal is moving. The determination of beams coming from antennas is performed in real time, which can lead to collisions.

6.13. Conclusion

The future belongs to small cells. The reasons for this are many, including perhaps the most important one coming from the forecast showing a sharp increase in rates. A high capacity could be obtained by very good frequency reuse. These small cells can be used everywhere, at home, in offices, in the streets, in shopping malls, airports, train stations, etc. One of the remaining challenges is to interconnect the core network. We will see in other solutions with mesh networks Chapter 12.

6.14. Bibliography

6.14.1. *Standards*

3GPP TS 23.003 v4.9.0 Numbering, addressing and identification.

3GPP TS 23.003 v8.6.0 Numbering, addressing and identification.

3GPP Requirements for Evolved UTRA (E-UTRA) and Evolved UTRAN (E-UTRAN) V7.3.0, 2006-03 (TR25.913).

3GPP Technical Specification Group Radio Access Network; Evolved Universal Terrestrial Radio Access (E-UTRA); User Equipment (UE) procedures in idle mode (TS36.304).

3GPP Technical Specification Group Radio Access Network; Evolved Universal Terrestrial Radio Access (E-UTRA) and Evolved Universal.

3GPP Technical Specification Group Services and System Aspects; 3G HNB and LTE HeNB OAM&P; Release 9, March 2010.

3GPP Technical Specification Group Services and System Aspects; Architecture aspects of Home NodeB and Home eNodeB; Release 9, TR 23.830; http://www.3gpp.org/.

3GPP – Technical Specification Group Services and System Aspects; Telecommunication management; Study of Self-Organizing Networks (SON) related Operations, Administration and Maintenance (OAM) for Home Node B (HNB); Release 9, TR 32.821; http://www.3gpp.org/

3GPP TR 33.820 v8.3.0., Technical Specification Group Service and System, Aspects; Security of H(e)NB; http://www.3gpp.org/ftp/Specs/archive/33 series/33.820/33820-830.zip.

TERRESTRIAL RADIO ACCESS NETWORK, (E-UTRAN) V8.5.0, 2008-05 (TS36.300).

TR-069 AMENDMENT 2, CPE WAN Management Protocol v1.1, Broadband Forum; http://www.broadbandforum.org/technical/download/TR-069 Amendment2.pdf

6.14.2. *Selected bibliography*

[BIL 10] BILOGREVIC I., JADLIWALA M., HUBAUX J.-P., "Security issues in next generation mobile networks: LTE and femtocells", *2nd International Femtocell Workshop*, Luton, UK, June 2010.

[CAL 03] CALLAWAY E.H., *Wireless Sensor Networks: Architectures and Protocols*, Auerbach Publications, 2003.

[CHE 06] CHEN Z.N., CHIA M.Y.W., *Broadband Planar Antennas: Design and Applications*, John Wiley & Sons, 2006.

[CHE 07] CHEN Z.N., *Antennas for Portable Devices*, John Wiley & Sons, 2007.

[DOR 02] DORMAN A., *The Essential Guide to Wireless Communications Applications*, Prentice Hall, 2002.

[FLI 01] FLICKENGER R., *Building Wireless Community Networks*, O'Reilly, 2001.

[GAS 02] GAST M.S., *802.11 Wireless Networks: The Definitive Guide*, O'Reilly, 2002.

[GOL 09] GOLAUP A., MUSTAPHA M., PATANAPONGPIBUL L.B., "Femtocell access control strategy in UMTS and LTE", *IEEE Communications Magazine*, vol. 47, no. 9, pp. 117–123, 2009.

[GSM 08] GSM ASSOCIATION, *Mobile Data Statistics*, 2008.

[HAC 03] HAC A., *Wireless Sensor Network Designs*, Wiley, 2003.

[HEL 03] HELTZEL P., *Complete Wireless Home Networking*, Prentice Hall, 2003.

[HOS 07] HOSSAIN E., LEUN K.K., *Wireless Mesh Networks: Architectures and Protocols*, 1st ed., Springer, 2007.

[KWA 08] KWAK H., LEE P., KIM Y. *et al.*, "Mobility management survey for home-eNB based 3GPP LTE systems", *Journal of Information Processing Systems*, vol. 4, no. 4 pp. 145–152, 2008.

[LEE 08] LEE S., "An enhanced IEEE 1588 time synchronization algorithm for asymmetric communication link using block burst transmission", *IEEE Communication Letters*, vol. 12, no. 9, pp. 687–689, 2008.

[LEE 09] LEE P., JEONG J., SAXENA N. *et al.*, "Dynamic reservation scheme of physical cell identity for 3GPP LTE femtocell systems", *Journal of Information Processing Systems*, vol. 5, no. 5, pp. 207–220, 2009.

[LOP 09] LÓPEZ-PÉREZ D., VALCARCE A., DE LA ROCHE G. *et al.*, "OFDMA femtocells: a roadmap on interference avoidance", *IEEE Communications Magazine*, vol. 47, no. 9, pp. 41–48, 2009.

[MAL 03] MALLICK M., *Mobile and Wireless Design Essentials*, Wiley, 2003.

[MAX 02] MAXIM M., POLLINO D., *Wireless Security*, McGraw-Hill, 2002.

[PAH 01] PAHLAVAN K., KRISHNAMURTHY P., *Principles of Wireless Networks: a Unified Approach*, Prentice Hall, 2001.

[PIC 07] PICOCHIP DESIGNS LTD, The Case for Home Base Stations, Technical White Paper v1.1, 2007.

[PRA 98] PRASAD R., *Universal Wireless Personal Communications*, Artech House, 1998.

[RAP 01] RAPPAPORT T.S., *Wireless Communications Principles and Practice*, Prentice Hall, 2001.

[SAY 01] SAYRE C.W., *Complete Wireless Design*, McGraw-Hill, 2001.

[SES 09] SESIA S., TOUFIK I., BARKER M., *LTE, The UMTS Long Term Evolution: From Theory to Practice*, Wiley, 2009.

[SIW 07] SIWIAK K., BAHREINI Y., *Radiowave Propagation and Antennas for Personal Communications*, 3rd ed., Artech House, 2007.

[STA 01] STALLINGS W., *Wireless Communications & Networks*, Prentice Hall, 2001.

[WAN 03] WANG X., POOR H.V., *Wireless Communication Systems: Advanced Techniques for Signal Reception*, Prentice Hall, 2003.

[ZHA 04] ZHAO F., GUIBAS L., *Wireless Sensor Networks*, Morgan Kaufmann Publishers, 2004.

6.14.3. *Websites*

FEMTO, Forum [Online], available at: http://www.femtoforum.org/femto/

IEEE 1588 Homepage, available at: http://ieee1588.nist.gov/

Internet World Statistics. [Online], available at: http://www.internet worldstats.com/

Next Generation Mobile Network [Online], available at: http://www. ngmn.org/

7

WPAN and WiGig

Wireless personal networks connect the equipment we carry on us with sufficient capacity to transmit voice, data and video. These are small networks, in the order of a few meters. Bluetooth has long been used, but is now rivaled by many proposals and products from standards set by the IEEE 802.15 Working Group. Currently, industrial bodies are pushing a new personal wireless generation network called Wireless Gigabit (WiGig) normalized under the name IEEE 802.11ad.

7.1. Wireless Personal Area Network

The IEEE 802.15 Working Group was set up in March 1999 to reflect on wireless networks with a range of 10 m, or Wireless Personal Area Network (WPAN), with the aim of connecting different mobile devices belonging to the same or multiple users. This type of network can connect a laptop, a cell phone, a PDA or any other device of this type. Three service groups were defined, A, B and C.

Group A uses the unlicensed spectrum band (2.4 GHz) targeting low cost implementation and use. The size of the cell around the point of emission is in the order of 1 m. Power consumption must be particularly low for the terminal to in last several months without electric charging. The selected transmission mode is connectionless.

The network must be able to work in parallel with an IEEE 802.11 network. In a single physical location, it may therefore be simultaneously a network of each type, both of which can possibly function in a degraded manner.

Group B shows the performance increase, with a medium access control (MAC) level up to a rate of at least 100 Mbps. The core network must be able to interconnect 16 or more machines and propose a quality of service, to authorize the operation of some applications, such as telephone speech, which requires a very strict quality of service. The range between the transmitter and receiver reaches 10 m, and the maximum time to connect to the network must not exceed 1 second. Finally, this network class must have bridges with other categories of 802.15 network.

Group C introduces important new features for individuals or businesses, such as communication security, video transmission and the possibility of roaming between wireless networks. To meet these objectives, industrial groups were set up, like Bluetooth SIG and WiMedia Alliance. The Bluetooth normalization group consists of more than 800 companies that have made an open specification for wireless connections between personal devices. Bluetooth is based on a radio link between two devices, while the WiMedia Alliance is interested in high-speed broadband connections over a short range. In fact, the standard may be any other standard with the arrival of WiFi type networks as part of personal networks.

7.2. IEEE 802.15

The IEEE 802.15 Working Group began by splitting into four subgroups:

– IEEE 802.15.1, for class C networks;

– IEEE 802.15.2 for dealing with interference problems with other networks using the 2.4 GHz band;

– IEEE 802.15.3 for class B networks;

– IEEE 802.15.4 for class A networks.

It was then greatly enlarged with the arrival of many groups or subworking groups, including the following:

– IEEE 802.15.3c (Task Group 3c), which provides a home network in the 57–64 GHz ranges, achieving transmission speeds of several gigabits per second;

– IEEE 802.15.5, which defines mesh networks, which can cover a larger geographic area than a personal network;

– IEEE 802.15.6, which deals with body area network (BAN), a network that people can carry with them for medical, experimental, sporting reasons, etc. The directions taken by the technical group point to lower power consumption and low frequencies with a range of 2–3 m. High security properties are of course essential in these BAN networks. Several proposals are currently under consideration: ultra wideband (UWB) specific WiFi, Bluetooth, etc;

– IEEE 802.15.7, which works in home networking technology with a visible light communication. This network uses equipment with LEDs and photodiodes, which can be found in neon lights, red traffic lights, light panels, etc. Transmission occurs between 400 and 800 THz and can provide speeds of several hundreds of megabits per second as an available bandwidth. It is possible to have extensions longer than a kilometer, for example, but with much lower flow rates (a few hundred kilobits per second).

The IEEE 802.15.1 Group turned to Bluetooth, presented in detail in the next section. This became the Bluetooth Special Interest Group (Bluetooth SIG). After several years, the IEEE announced in 2005 the abandonment of Bluetooth technology in favor of UWB technology.

The IEEE 802.15.3 Working Group is attempting broadband and UWB standardization. This interface was re-covered by the WiMedia Alliance and the Bluetooth SIG. After many negotiations, issues and technical discussions on patents, the Bluetooth SIG abandoned the UWB solution.

7.3. Bluetooth

The Bluetooth SIG defines Bluetooth specifications. This group was initially formed by Ericsson, IBM, Intel, Nokia and Toshiba, and has been joined by more than two 2,500 companies. Bluetooth is a low-cost technology, due to its strong integration on a single 9 mm x 9 mm chip. The frequencies used are between 2,400 and 2,483.5 MHz. We find the same frequency range in most wireless networks used in private environments, whether personal or business. This band requires no license.

It was long thought that UWB technology could succeed or be integrated into Bluetooth: this was the view of the IEEE 802.15.1. However, the power consumption of this solution has proved quite prohibitive. That is why WiFi is likely to take a prominent place in personal broadband networks, either by Bluetooth extensions or purely WiFi networks. The succession of Bluetooth SIG standards is as follows:

– Bluetooth 1.0: this had many flaws, including the fact that the marketed products could be incompatible;

– Bluetooth 1.1: first real standard ratified by IEEE 802.15.1 in 2002;

– Bluetooth 1.2 brings many improvements, such as greater environmental discovery speed and a transmission capacity of 721 Kbps. This is the standard ratified by the IEEE 802.15.1 Working Group in 2005, which we will detail below;

– Bluetooth 2.0 + EDR: greatly increases the throughput by an enhanced data rate solution. The new value of the flow is 3 Mbps, but the actual speed, is well below 2 Mbps. This increase is due to a more efficient modulation, using both Gaussian Frequency Key Shifting and phase shift keying carrying two or three bits per band;

– Bluetooth 2.1 + EDR approved in 2007 by the Bluetooth SIG, is an interim release improving security and pairing terminals;

– Bluetooth 3.0 + HS was adopted in 2009. Since the division of physical media and transmission technologies is not efficient in Bluetooth, the idea is to add a new technique for the transmission other than the classical Bluetooth solution. Bluetooth low energy technology is used when the network is idle, and WiFi technology at the time of transmission;

– Bluetooth 4.0: the 2010 standard intends to have as low a consumption as possible and to collate the different Bluetooth solutions. Nokia, which had researched low consumption Bluetooth, proposed a new product, called Wibree. Rather than standardize this new product, Nokia joined forces in 2007 to incorporate the Bluetooth SIG advanced Wibree into Bluetooth. The provisional name of the new generation, Bluetooth Ultra Low Power, was abandoned in April 2010 to become Bluetooth Low Energy. Bluetooth 4.0 brings together in one standard basic Bluetooth 2.1, Bluetooth 3.0 + HS, with WiFi for transmission and Bluetooth Low Energy.

Several connection diagrams have been defined by the standard setters. The first corresponds to a single network, called piconet, which can support up to eight terminals, with a master and eight slaves. The master terminal handles communications with the different slaves. Communication between two slave terminals necessarily passes through the master terminal. In one piconet, all terminals use the same hopping sequence.

Another communication scheme interconnects piconets to form a scatternet. As communications are always as master–slave, the master of a piconet can be a slave to the master of another piconet. For its part, a slave can be the slave of several masters. A slave can temporarily detach a master to cling to another piconet and then back to the first master, once its communication with the second is completed.

Figure 7.1 shows the Bluetooth terminal connections in which a master of a piconet is a slave of another piconet master and a slave is a slave to two masters. Overall, three piconets are interconnected by a master to form a scatternet.

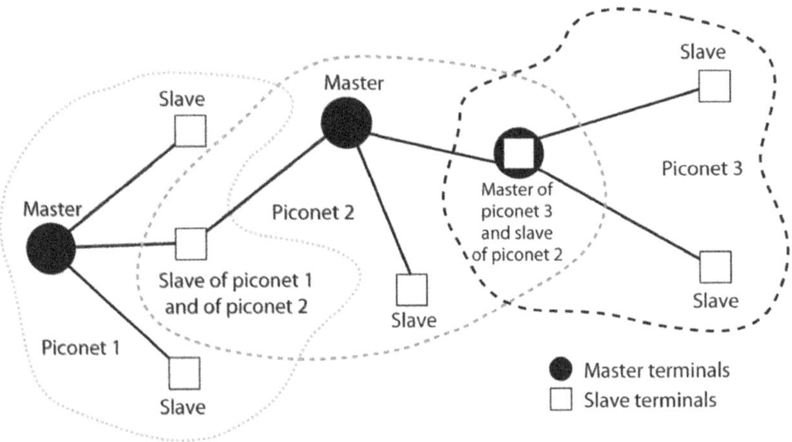

Figure 7.1. *Bluetooth terminal connection diagram*

Communication within a piconet can reach close to 1 Mbps. As there may be up to eight terminals, actual speed rapidly decreases depending on the number of terminals connected in a single picocell. A master may, however, increase its communication speed by working with two slaves using different frequencies.

A Bluetooth connection between two machines can reach a rate of 433.9 Kbps bidirectional communication (full duplex). The flow rates are equal to 723.2 Kbps in one direction and 57.6 Kbps in the other with unbalanced communication. Communications may be of two types: asynchronous or synchronous. Synchronous communication, or synchronous connection-oriented (SCO) link allows a synchronous data rate of 64 Kbps. This type of connection allows the passage of telephone speech with a service guarantee. asynchronous communication, or asynchronous connection-less link allows asynchronous traffic with more or less protection. The flow rate can reach 723.2 Kbps.

Several categories of communications can be defined over a Bluetooth connection: a single asynchronous communication, three simultaneous communications SCO or SCO with a symmetrical LCD 433.9 Kbps. This gives a total of the link throughput of almost 1 Mbps in the latter case. A slave device can only support a maximum of two SCO channels from two separate terminals. More specifically, time is sliced, into slots, at a rate of 1,600 slots per second. A slot is therefore 625 µs long, as shown in Figure 7.2. A terminal uses a frequency on a slot and then by frequency hopping (frequency hop) it changes frequency to the next slot and so on.

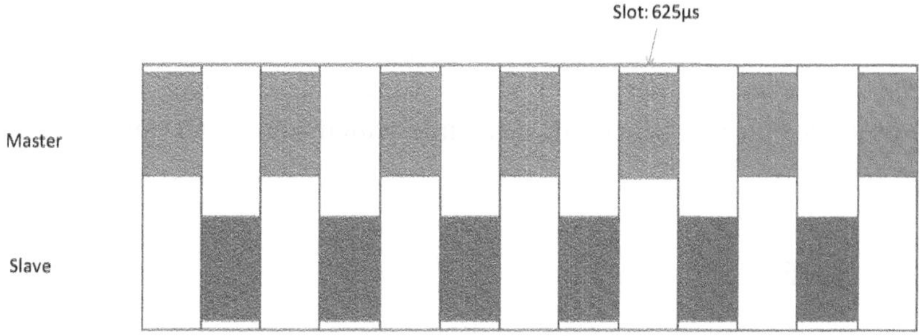

Figure 7.2. *Cutting slots*

A Bluetooth client cyclically uses all the frequency bands. Customers within the same piconet have the same sequence of frequency hopping, and when a new terminal wants to connect, it must begin by recognizing all frequency hopping in order to respect it. Communication is exercised through a series of packets. A packet stands on a slot. A series of packets may extend, over three or five slots (see Figure 7.3). Frequency hopping takes place at the end of the communication of a packet.

As indicated above, Bluetooth allows the production of small personal networks of a few square meters, the piconets. The terminals are connected by means of a master. Transmission powers up to 100 mW (milliwatt), which allows a program to cover several tens of meters. It is possible to reduce this power to 2.5 mW and even 1 mW to reach a range of a few meters.

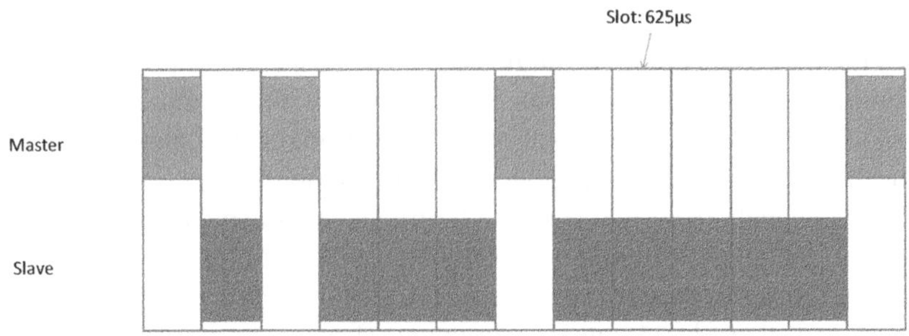

Figure 7.3. *Transmission on several slots*

At an output of 100 mW, a battery can hold out long enough, using the power-saving options. For this reason, low power states were introduced in the Bluetooth standard, which allows an autonomy of several days.

7.4. UWB

The MultiBand OFDM Alliance (MBOA) standardization group has defined the basic elements in an ecosystem that would offer consumers a wide range of products for the management and control of home networks. The UWB radio layer and convergence platform together form the basic transport mechanism for different applications, including IEEE 1394 (FireWire), and the framework from the Digital Living Network Alliance consortium.

MBOA, established in 2003 and formalized in (SIG) in 2004, was established to promote the production of a global standard for wireless UWB solutions. Its 200 members included semiconductor companies, scientific computing, consumer electronics and mobile devices. The characteristics of the physical layer and media access controller are finalized and available to MBOA member companies. MBOA is part of another more important alliance, the WiMedia Alliance. Unfortunately, the UWB technology for power consumption reasons did not achieve the expected success. Several network standards, like Bluetooth, abandoned the idea of taking UWB as their physical level. After many discussions, the WiMedia Alliance was dissolved in 2009, pushing the Wireless USB Promoter Group (often referred to as Wireless USB and even WUSB) to promote the basic product of the WiMedia Alliance.

The WiMedia Alliance is an industrial group that has contributed significantly to the development of the UWB concept, particularly in the form of software-level protocol layers. This alliance aims to promote the connectivity and interoperability of UWB products. The common platform illustrated in Figure 7.4 enables the multiple applications to work together on a common radio interface.

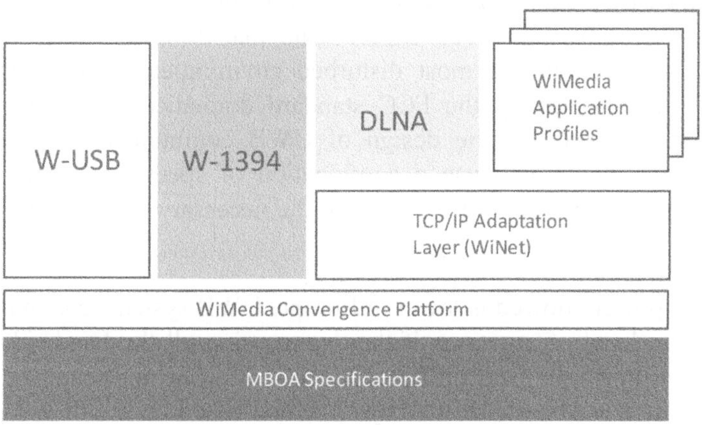

Figure 7.4. *Platform WiMedia Alliance*

The combination of the proposed radio interface using the multiband OFDM and architecture of the WiMedia platform allows convergence for the implementation of wireless versions of UWB: WUSB, 1394, and WiNet DLNA.

UWB defined, according to the US Federal Communications Commission (FCC), any radio technology spanning part of the spectrum covering more than 20% of the center frequency or at least

500 MHz. Identifying the benefits of new products that could incorporate this technology to benefit public safety applications, corporate and limited energy consumption, in 2002 the FCC allocated a part of the usable radio unlicensed spectrum ranging from 3.1 to 10.6 GHz. An additional part of the spectrum is available for medical, scientific and fire rescue.

Rather than requiring a UWB interface using all the available bandwidth, the FCC has set a minimum bandwidth of 500 MHz to 10 dB. This minimum bandwidth and other requirements imposed by the FCC protect the transmission of signals from other users of this part of the spectrum. While transmitting under the noise cannot disturb other band users, however, these other users could disrupt UWB. For example, a UWB connection at an airport would be disrupted by radars and other equipment using the UWB band. It has been shown by numerous experiments that 500 MHz was always available, even in the most disturbed environments. However, the flexibility allowed by the FCC standard dramatically increases the available options for the design of UWB communication systems. Designers are free to use a combination of spectrum subbands to optimize the operation of the system, the necessary power and design complexity.

The power allowed is lower in Europe. UWB systems can maintain the same level of power as if they were using all the bandwidth, by inserting symbols in the different subbands. In a multiband system, information can be transmitted by a method based on traditional pulse-based carrier or more advanced multicarrier techniques.

Systems based on the pulses on the basis of carrier transmit signals by modulating the pulse phase. This solution comes from proven technology, which relies on a simple design of the transmitter but is not, however, without drawbacks. It is difficult to gather enough energy to transmit the signal in a typical use environment, with lots of reflective surfaces, using only a simple RF chain. The limitations of switching times can be very strict, to the transmitter as the receiver.

The Multiband OFDM (MB-OFDM) simultaneously transmits multiple carriers spaced at precise frequencies. The use of fast Fourier transform algorithms guarantees high-energy efficiency in the case of multipath, while only slightly increasing the complexity of the transmitter. The benefits of this solution include MB-OFDM flexibility, high spectral elasticity and a high resistance to radio interference and effects from multipath.

OFDM modulation techniques have been successfully applied to many communication systems marketed as high performance, including WiFi IEEE 802.11a/g, 802.16a WiMAX, ADSL and HomePlug standards. Based on CMOS technology, the 3.1–4.8 GHz spectrum is considered optimal for initial deployments. The upper limit is to prevent interference with the unlicensed national information infrastructure band, is located the 802.11a radio interface, and simplify the design of radio and analog circuits. The 3.1–4.8 GHz frequency band is sufficient for three 500 MHz subbands, as shown in Figure 7.5.

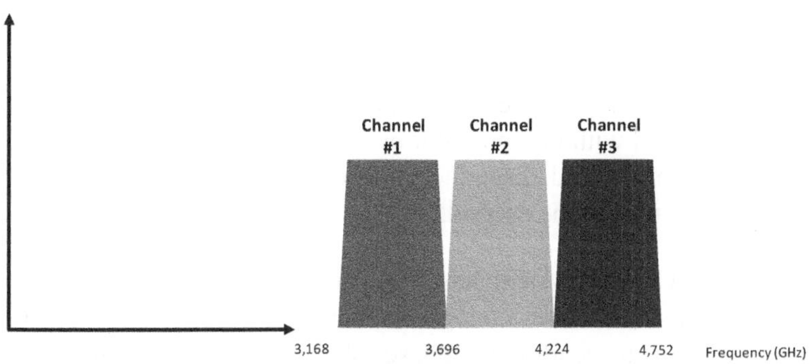

Figure 7.5. *Allocation of frequencies for MB-OFDM*

The IEEE 802.15.3 Working Group developed two solutions, one on the 2.4 GHz band, which will reach a speed of 54 Mbps, and a second using the entire bandwidth between 3.1 and 10.7 GHz, but at very low power, below the ambient noise. Thus, this solution does not interfere with civilian and military applications using these bands. The spectrum covered by this technology is shown in Figure 7.6.

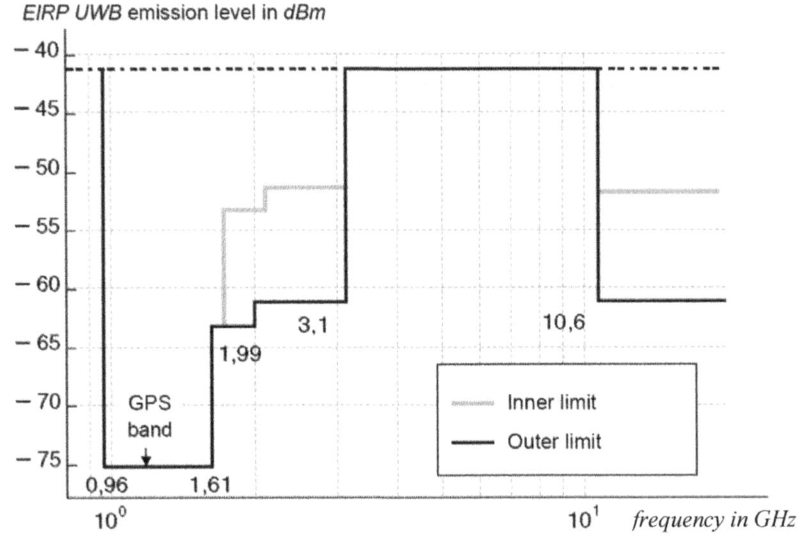

Figure 7.6. *Part of the spectrum can be used by the UWB*

Despite the very low power used, the bandwidth of over 7 GHz provides a speed between 120 and 480 Mbps depending on external perturbations. One of the properties of UWB is it can support communications with equipment moving at relatively slow speed. The goal is to connect and disconnect the equipment in extremely short times, in the order in the second.

The MB-OFDM system has been specifically designed to be as simple as possible. One analog receiver chain restricts the symbols transmitted in a quadrature phase shift keying modulation. The relatively large spacing between the carriers makes it possible to reduce noise on the circuit and to improve the robustness to synchronization errors. Battery life is a critical factor for users. MB-OFDM Access is able to last several hours under typical conditions before recharging is required.

UWB technology is designed to ship the elements necessary assured to security at all times. Privacy mechanisms are implemented at several protocol levels to ensure the robustness needed for wireless environments while remaining transparent to users. The experience gained with the growth of WiFi networks and Bluetooth has guided the choices made for security architecture. Most solutions implemented are identical to those promoted by the IEEE 802.11i standard that we will detail in the next chapter.

Communication security is provided by authentication and encryption. Authentication uses the Extensible Authentication Protocol–Transport Layer Security. The use of the advanced encryption standard with 128 bit key must guarantee the safe transport of information.

7.5. WiGig

The Wireless Gigabit Alliance was formed in 2009 to create a wireless network in the 60 GHz band capable of exceeding speeds of one gigabit per second. WiGig, known as 802.11ad, is known as "microwave WiFi". Given the choice of the frequency band, this network cannot be a personal network since the highest frequency used does not allow the signal to traverse obstacles. The final WiGig specification was announced in December 2009. In May 2010, WiGig released its version 1.0 specifications, opened its adopter program and entered a cooperation agreement with the WiFi Alliance to develop free WiFi technologies. In June 2011, WiGig announced the launch of its certification for version 1.1 specifications.

The Wireless Gigabit Alliance may in the longer term provide specifications in classic wireless frequencies with a transfer capacity of more than 1 Gbps in the 2.4 GHz band. The maximum WiGig capacity is 7 Gbps, but the actual flow rate depends on the environment, i.e. mainly on the removal of the access point and interference.

The IEEE Working Group performs 802.11ad WiGig standardization work in the 60 GHz band, which has sufficient free tapes to reach the rate of 7 Gbpspic rate capacity. This working group fits into a module WiFi 802.11g or n WiFi, and a new generation of the 60 GHz band. The idea is to look for devices that need to communicate using the WiFi standard and hand over to the high-frequency part to get the advertised speeds.

The WiGig standard allows devices to communicate wirelessly at speeds that are measured in multi-gigabits. WiGig devices become compatible with other WiFi solutions by adopting a tri-band technology, which works in the 2.4, 5 and 60 GHz bands. It offers data transfer rates up to 7 Gbps, roughly as much as transmission with eight 802.11ac antennas, and nearly five times more than the highest WiFi 802.11n rate while maintaining compatibility with existing WiFi devices. However, the 7 Gbps rate uses the 60 GHz band, which does not pass through walls; it is a line-of-sight technology. This solution is primarily intended for indoor areas, because the frequencies are strongly dampened by water droplets or dust outside the buildings. However, outdoor communications use the 60 GHz over distances of a kilometer in direct view. While roaming, tri-band WiGig uses classic bands but can better propagate through walls. If necessary, the tri-band shifts into high gear if direct view is possible.

WiGig has the following characteristics:

– it supports data rates up to 7 Gbps;

– it complements the 802.11 MAC layer, while remaining compatible with various IEEE 802.11 standards;

– it has a specific physical layer ensuring interoperability with WiGig equipment allowing them to reach speeds of several gigabits per second;

– it adaptation layers that support specific system interfaces, such as data buses for PC peripherals and display interfaces for high-definition television screens and projectors;

– direction through the physical layer of beams allowing communication at distances of at least 10 m. The beam can move within the coverage area through a transmission phase change;

– standard WiGig offering the same kind of security as WiFi networks.

The WiFi Alliance and WiGig Alliance created an agreement for wireless multi-gigabit networks. This cooperation has helped to produce specifications for the development of a WiFi product certification program in the 60 GHz frequency band.

The WiGig Alliance and the Video Electronics Standards Association (VESA) agreed to develop a new generation of wireless display. VESA and the WiGig Alliance share the WiGig technology specifications to develop functions such as multi-gigabit wireless DisplayPort without creating a certification program for wireless DisplayPort products.

The WiGig Alliance has agreements for using the high definition multimedia interface (HDMI). The agreement provides for mapping displays for many applications such as wireless transmission of video, whether compressed or not.

The channels used in WiGig are given in Table 7.1.

Channel	*Center [GHz]*	*Min [GHz]*	*Max [GHz]*	*BW [GHz]*
1	58.32	57.24	59.4	2.16
2	60.48	59.4	61.56	2.16
3	62.64	61.56	63.72	2.16
4	64.8	63.72	65.88	2.16

Table 7.1. *WiGig channels*

Depending on the quality of the communication, the different data rates reached by WiGig are given in Table 7.2.

Phy Rate (Mbps)
693
866.25
1386
1732.5
2079
2772
3465
4158
4504.5
5197.5
6237
6756.75

Table 7.2. *Data rates reached by WiGig*

7.6. WirelesssHD

The basic idea of the wireless HD group was to develop an air interface for HDMI. The product developed reaches 25 Gbps. The consortium currently has more than 50 members. The WirelessHD specification is based on a channel 7 GHz wide in the 60 GHz frequency band. This product enables digital transmission with or without compression of audio signals and high-definition HDMI-type Wireless video. The first generation reached the 4 Gbps throughput, but the core technology allows theoretical speeds of up to 25 Gbps (against 10.2 Gbps for HDMI 1.3 and 21.6 Gbps DisplayPort 1.2). Version 1.1 of the specification further increases the maximum throughput to 28 Gbps. This version is compatible with common 3D formats, 4K resolution, WPAN networks and HDCP 2.0 content protection.

The 60 GHz band requires a direct line of site between the transmitter and the receiver. However, the WirelessHD specification restricts this limitation by using beams at the receiver and transmitter to increase the effective radiated power of the signal and find the best path using wall reflections. The first product is to be used inside a part in point-to-item without being in direct line of sight, or non-line-of-sight at distances of up to 10 meters.

WirelessHD specification has security features encrypting content via the Digital Transmission Content Protocol and for performing network management. A standard remote control allows users to control WirelessHD devices and choose source of information to be displayed on the remote screen.

7.7. Conclusion

Personal networks are a wide range of solutions ranging from very high speed to low speed with extremely diverse energy consumptions.

Two main aims are: broadband and low-energy consumption. Bluetooth is not a broadband solution and UWB consumes lots of energy. Therefore, a new generation of WiFi is entering the market: WiGig.

7.8. Bibliography

[ALL 06] ALLEN B., DOHLER M., OKON E.E. *et al., Ultra-Wideband Antennas and Propagation for Communications, Radar and Imaging*, Wiley, 2006.

[ARS 06] ARSLAN H., CHEN Z.N., DI BENEDETTO M.-G., *Ultra Wideband Wireless Communication*, John Wiley & Sons, 2006.

[BAK 02] BAKKER D., MCMICHAEL GILSTER D., GILS R., *Bluetooth End to End*, 1st ed., Wiley, 2002.

[FAR 08] FARAHANI S., *ZigBee Wireless Networks and Transceivers*, Newnes, 2008.

[GHA 07] GHAVAMI M., MICHAEL L., KOHNO R., *Ultra Wideband Signals and Systems in Communication Engineering*, 2nd ed., Wiley, 2007.

[HUA 07] HUANG A.S., RUDOLP L., *Bluetooth Essentials for Programmers*, 1st ed., Cambridge University Press, 2007.

[KUB 07] KUBAN P.A., *A Controller Area Network Gateway to ZigBee – A Proposition of an Architecture to Extend CAN*, VDM Verlag, 2007.

[KUO 08] KUORILEHTO M., KOHVAKKA M., SUHONEN J. *et al.*, *Ultra-Low Energy Wireless Sensor Networks in Practice: Theory, Realization and Deployment*, Wiley, 2008.

[LAB 07] LABIOD H., AFIFI H., DE SAN C., *Wi-Fi, Bluetooth, Zigbee and WiMax*, 1ST ED., Springer, 2007.

[OPP 04] OPPERMANN I., HÄMÄLÄINEN M., IINATTI J., *UWB: Theory and Applications*, Wiley, 2004.

[SIW 04] SIWIAK K., MCKEOWN D., *Ultra-wideband Radio Technology*, Wiley, 2004.

[WOO 08] WOOD S., AIELLO R., *Essentials of UWB*, 1st ed., Cambridge University Press, 2008.

[WU 07] WU S.-L., TSENG Y.-C., *Wireless Ad Hoc Networking: Personal-Area, Local-Area, and the Sensory-Area Networks*, 1st ed., Auerbach, 2007.

8

WLAN and WiFi

The IEEE 802.11 standard has resulted in four types of wireless networks, which have different peak rates: 11 Mbps, 54 Mbps, 600 Mbps and 2 Gbps. The first is based on IEEE 802.11b, the second on IEEE 802.11a and g, the third on IEEE 802.11n and the fourth on beamforming. These four generations will be succeeded by three new generations with IEEE 802.11af, which could reach a global capacity of 10 Gbps, IEEE 802.11h to connect things on a large range, and IEEE 802.11ax with eight beamforming waves and several gigabit per second peak rate capacity.

For the existing types, the selected frequencies are in the range of 2.4 or 5.15 GHz. In these Local Area Networking (LAN) solutions using radio technology, communications can be done either directly from station to station, but without a station the packet can be relayed to another terminal station, or via an antenna called the access point (AP).

Access to the physical media is performed via a Medium Access Control (MAC) protocol internal to the MAC layer for all types of WiFi network. Many options can make the implementation rather complex. The MAC protocol is based on the Carrier Sense Multiple Access/Collision Detection (CSMA) access technique, already used in Ethernet networks. The difference between the wireless protocol and the terrestrial protocol comes from the way collisions are detected. In the cable version, collisions are detected by listening to the carrier. When two stations want to transmit at the same time a third one is in the process of transmitting a frame then this leads automatically to a collision. In the wireless case, the access protocol avoids collisions by forcing both stations to wait before transmitting. Since the difference between waiting times is greater than the propagation time on the transmission medium, the station which has the longest waiting time is the physical media already occupied, thereby avoiding the collision. This new technique is called CSMA/ Collision Avoidance (CA).

To avoid collisions, each station has a timer with a specific value. A station can transmit when discovering an empty channel by listening to the carrier. The risk of a collision occurring is extremely low, since the probability that two stations start their emission in the same microseconds is almost zero. However, when several stations begin to listen during a frame transmission and continue to listen, the collision becomes unavoidable. To prevent collision, the stations must wait before transmitting to separate their respective transmission instants. A first timer is added for a very small period of time allowing the receiver to immediately send an acknowledgment. A second timer is used to give high priority to real-time application. Finally, the longest timer corresponds to asynchronous packets with no real-time constraints.

8.1. IEEE 802.11

WiFi networks come from the IEEE 802.11 standard, which defines a cellular architecture. A group of terminals equipped with 802.11 network interface cards can establish communications through an antenna or directly with each other. They then form a Basic Service Set (BSS), not to be confused with the Base Station Subsystem (BSS) of the GSM networks. The area occupied by the terminal of a BSS may be a basic set area or cell.

As shown in Figure 8.1, IEEE 802.11 standard offers two modes of operation: the infrastructure mode and the ad-hoc mode. The infrastructure mode is set to provide different stations of specific services over a coverage area determined by network size. Infrastructure networks are established using APs, which play the role of a base station for a BSS.

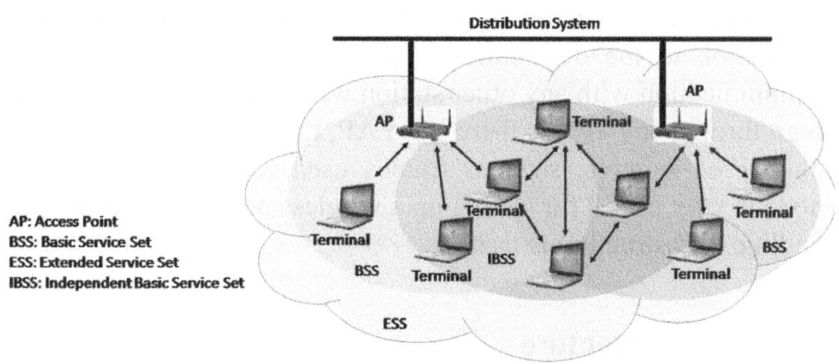

Figure 8.1. *IEEE 802.11 mode of operation*

When the network is composed of several BSSs, each AP is connected to a distribution system. The distribution system generally corresponds to a wired Ethernet network. A group of BSS interconnected by a distribution system is an extented service set, which is not very different from a mobile radio network subsystem.

The distribution system is responsible for transferring packets between BSSs that are different from the same ESS. In the specifications of the standard, the distribution system is implemented independently of the radio structure of the wireless system. That is why the distribution system almost always corresponds to an Ethernet network. Nothing would prevent US using other types of LAN. Another solution is to use the WiFi network itself, which gives a mesh network.

The ESS can provide different mobile stations with an access gateway to a fixed network, such as the Internet. This gateway connects the 802.11 network to another network. If this network corresponds to an IEEE 802.x environment, the gateway incorporates functions similar to those of a bridge.

A network in ad-hoc mode is a group of terminals forming an Independent Basic Service Set (IBSS), whose role is to allow stations to communicate without using any infrastructure, such as an AP or connection to the distribution system. Each station can establish a communication with any other station within the IBSS, without having to go through an AP. As there is no AP, the stations will incorporate a number of features, such as frames used for synchronization. This mode is very useful for setting up a wireless network when wireless or landline infrastructure is lacking.

8.2. WiFi architecture

IEEE 802.11 covers the first two layers of the open systems interconnection reference model. One of its key features is that it defines a MAC layer common to all physical layers. In this way, future physical layers may be added without having to change the MAC layer.

8.2.1. *Physical layer*

The physical layer has the role of correctly transporting 0 or 1 signals that the sender wants to send to the receiver. The layer is divided into two sublayers: Physical Layer Convergence Protocol (PLCP) and Physical Medium Dependent (PMD).

The PMD takes care of encoding the data, while the PLCP sublayer supports listening to the media. It provides a clear channel assessment which is used by the MAC layer to determine whether the medium is busy or not.

The CSMA/CA collision avoidance technique uses acknowledgment frames or ACK. IEEE 802.11 defines four different physical layers:

– frequency hopping spread spectrum (FHSS);

– direct sequence spread spectrum (DSSS);

– infrared;

– orthogonal frequency division multiplexing (OFDM).

FHSS and DSSS use the 2.4 GHz Industrial, Scientific and Medical (ISM) band, an unlicensed band. Infrared is only used in cases where the distances between the different stations are low.

The fourth physical layer has been defined in the 5.2 GHz band. With OFDM coding, bit rates of 54 Mbps have been achieved IEEE. 802.11 is the first standard to use an OFDM coding for a packet type of communication. This technology has so far been used for continuous data transmission systems, such as digital video broadcasting and digital audio broadcasting.

To be received correctly, the transmission cannot exceed 50 m in an office environment, 500 m without obstacle and several kilometers with an external directional antenna. When a wall has to be traversed, this distance is often much shorter.

8.2.2. *Data link layer*

The data link layer consists essentially of two sublayers: logical link control (LLC) and MAC. The LLC layer uses the same properties as the 802.2 LLC layer. It is thereby possible to connect a WLAN to any other local network belonging to the corresponding IEEE standard. The MAC layer, in turn, is specific for 802.11.

The role of the 802.11 MAC layer is similar to the 802.3 MAC layer of the Ethernet terrestrial network, since the terminals listen to the carrier before transmitting. If the carrier is free, the terminal transmits, otherwise it waits. The 802.11 MAC layer integrates many features that are not found in the wired version. The features required to achieve access over a radio interface are:

– support allocation procedures;

– packet addressing;

– frame formatting;

– cyclic redundancy check;

– fragmentation-reassembly.

One of the standard features is that it basically defines two different access methods at the MAC layer. The first is the distributed coordination function (DCF), which corresponds to a similar access method to traditional networks supporting best-effort. The DCF is designed to support asynchronous data transport, in which all users who want to transmit data have an equal opportunity to access the physical medium. The second access method is point coordination function (PCF). This solution is based on interrogation of the different terminals, or polling. Polling is controlled by the AP. The PCF method is designed primarily for the transmission of sensitive data, requiring management of time used for real-time applications, such as voice or video. A network in ad-hoc mode only uses DCF, while an infrastructure network with the AP uses both DCF and PCF techniques.

8.2.3. *Access techniques*

DCF is the general access technique used to allow asynchronous best-effort data transfers. According to the standard, all stations must support this technique. DCF relies on the CSMA/CA access method.

In the Ethernet, the protocol that implements the CSMA/CD controls each stations' access to the carrier. The protocol detects and processes collisions that occur when two or more stations are transmitting simultaneously. In WiFi networks, collision detection is not possible due to the radio signal. To detect a collision, a station must be able to listen and transmit at the same time. In radio systems, transmission covers reception signals on the same frequency and does not allow the station to hear the collision: the radio links are never full-duplex. As a node can listen to its own transmission, if a collision occurs, the station continues to transmit the entire frame, resulting in a loss of network performance. WiFi access technique must take account of this phenomenon.

8.2.4. *The CSMA/CA protocol*

The CSMA/CA access method with collision avoidance uses acknowledgment frames or Acknowledgement (ACK). An ACK is sent from the destination station to confirm that the data are received correctly.

Media access is controlled by the use of inter-frame spacing (IFS), which corresponds to the time interval between the transmission of two frames. IFS intervals are periods of inactivity on the transmission medium. Different values of the IFS period are calculated by the physical layer. The standard defines three types of IFS (SiFS):

– short inter-frame spacing: the smallest IFS is used to separate transmissions within the same dialogue (sending data, ACK, etc.). There is always one station to transmit at this time, thus having priority over all other stations. The value of the short inter-frame spacing is 28 µs;

– point coordination function IFS is used by the AP to access the transmission media with priority. It has the same value as short inter-frame spacing with an additional time, or timeslot, defined in the back-off algorithm, of 78 µs;

– distributed coordination function IFS (or DIFS) is used when a station wants to start a new transmission. It has the same value as point coordination function IFS with an added time of 128 µs.

The terminals with the same BSS can listen to the activity of all other stations. When a station sends a frame, the other stations listen to avoid a collision, and update the Network Allocation Vector (NAV) timer to delay all planned transmissions. The NAV is calculated in relation to information located in the time-to-live field, or TTL, contained in frames that are sent. The other stations have the ability to transmit only after the end of the NAV.

During a dialogue between two stations, the NAV is calculated on the TTL field of different frames that are sent (data, ACK, etc.). The NAV is actually a timer that determines the timing at which the frame can successfully be transmitted. A source station wishing to transmit data listens to the media. If no activity is detected for a period of time corresponding to a DIFS, it transmits its data immediately. If the media is still busy, it continues to listen until it is free. When the media becomes available, the station still delays the transmission using the back-off algorithm before transmitting its data.

If the sent data are received correctly, the destination station waits for a time equivalent to a SIFS and sends an ACK to confirm good reception. If the ACK is not detected by the sending station or if the data are not successfully received or if the ACK is not received correctly, it is assumed that a collision occurred, and the frame has to be retransmitted. When the source station transmits its data, the other stations must update their NAV, including the transmission time of the data frame, the SIFS and the ACK.

Figure 8.2 shows the process of transmitting frames from a transmitter. This process includes the different expectations we just detailed.

The back-off algorithm solves the problem of media access when many stations want to transmit data simultaneously. In WiFi, the time is sliced, each slice corresponding to a timeslot. Unlike the timeslot used in aloha, which corresponds to the minimum transmission time of a frame, the timeslot used in WiFi is a little smaller. It is used to define the IFS intervals and timers for the different stations. Its implementation is different for each physical layer.

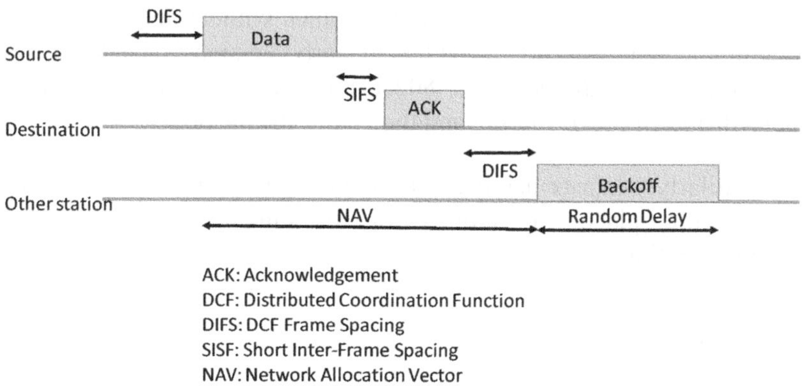

Figure 8.2. *Frame transmission process*

As soon as a collision occurs, a station calculates the value of a timer, called the back-off timer, corresponding to a certain number of timeslots. When the media is free, stations decrement their timer until the medium is busy or the timer reaches 0. If the timer has not reached 0 and the support is again busy, the station stops the timer. Once the timer reaches 0, the station transmits its frame. If two or more stations reach the value 0 at the same time, a collision occurs, and each station must generate a new timer. For each retransmission attempt, the timer is based on a random value M defined within an interval which increases as follows:

$$0 \leq M < 2^{i+1}$$

where *i* is the number of consecutive times a station tried to send a frame. This random value is then multiplied by the timeslot to get the timer.

With this algorithm, the stations have an equal probability of accessing the media. The only disadvantage is it does not ensure minimal delay and thus complicates the management of real-time applications such as voice or video.

The transition from one cell to another without interrupting the communication was not provided in early versions. Handover was introduced through the IEEE 802.11f technique with a quite large delay between the beginning and the end of the handover.

Similarly, security has been strengthened to prevent one customer taking the place of another or listening to another user's communications.

8.2.5. *Handovers*

In wireless networks, handovers occur when a terminal wishes to move from one cell to another without interrupting communication. They are using the same solution as in mobile telephony, with some nuances. In wireless networks, the handover takes place between two data transmissions and not in the middle of a dialogue.

The standard does not provide the handover mechanism in itself but sets some rules, such as synchronization, passive and active eavesdropping or association and reassociation, which allow stations to select the AP to which they want to associate.

When terminals move (change cell or in power-saving mode), they must remain synchronized to communicate. At a BSS, stations synchronize their clocks with the clock of the AP. To maintain synchronization, the AP periodically sends beacon frames, which contain the clock value of the AP. Upon reception of these frames, stations update their clock to stay synchronized with the AP.

After switching on or return from a sleep mode or a handover – when a terminal wants to access a BSS or ESS controlled by one or more APs – this terminal chooses an AP on which it connects, according to a number of criteria such as signal strength, packet error rate and network load. If the power of the AP is weak, the station seeks a more appropriate AP. This search for the best AP is performed by listening to the media. The latter can be done in two different ways, passive or active, according to criteria such as performance or power consumption:

– passive listening: the station waits to receive a beacon frame from the AP;

– active listening: when the station has found the most appropriate AP, it sends an association request through a frame called the Probe Request Frame and waits for the AP to respond.

As soon as the terminal is accepted by the AP, it can be adjusted in the most appropriate radio channel. Periodically, the terminal monitors all network channels to judge whether an AP has better performance. If this is the case, it is associated with the new AP and adjusts its radio channel accordingly.

Reassociations are made when a station is physically moved from its original AP, resulting in a decrease in signal strength. In other cases, reassociations are due to changes in the characteristics of the radio environment or due to excess network traffic on the original AP. In this case, the standard provides a load balancing function, which allows us to distribute the load effectively within the BSS or ESS and so avoid reassociations.

8.2.6. *Security*

In wireless networks, the radio channel is shared. All transmitted information can be intercepted. To enable wireless networks to have traffic as secure as in fixed networks, the 802.11 Working Group has developed the Wired Equivalent Privacy (WEP), whose mechanisms are based on data encryption and station authentication. According to the standard, WEP is optional. Terminals and APs are not required to implement it. As we will see, security is not guaranteed with WEP, and an attacker can break the encryption keys without much difficulty. The WiFi Alliance has developed a second protection mode, WiFi Protected Access, which solves these problems. Finally, the 802.11

Working Group created a specific subgroup, IEEE 802.11i, which offered a sustainable solution, standardized in June 2004: WPA2.

Before presenting these three security protocols, recall two elementary protection rules:

– hide network name, or Service Set Identified (SSID), so that a user does not see the network and therefore cannot connect to it. This security measure is unfortunately provisional. If an attacker listens to the network long enough, it will see the name of the network since a user who wants to connect must give this SSID;

– only allow communication controlled by a MAC address list, or access control list. This will not provide access to stations whose MAC address is not specified in the list.

8.2.7. *Wired Equivalent Privacy*

As explained above, the trust and authentication solution marketed with WiFi equipment is WEP. Frames transmitted over wireless networks are protected by encryption. Only decryption with the static WEP key shared between the terminal and the network is allowed. This key is obtained by concatenation of a secret key of 40 or 104 bits and an initialization vector of 24 bits. The size of the final key is 64 or 128 bits.

From the key obtained, the Ron's Code 4 (RC4) algorithm performs the encryption of data in stream mode (stream cipher). A RC4 key has a length of between 8 and 2,048 bits. The key is placed in a pseudorandom number generator. This generator determines a pseudo random sequence of bytes, or keystream. This series of bytes, called Ksi, is used to encrypt a plaintext message (M) using a conventional protocol, performing an exclusive-or (XOR) between Ksi and Mi:

$$Ci = Ksi \oplus Mi.$$

The Mi message consists of data that are concatenated with their integrity check value. The encrypted frame is then sent with his IV. The IV is an index that is used to recover the keystream and thus decrypt the data. The WEP encryption process is shown in Figure 8.3.

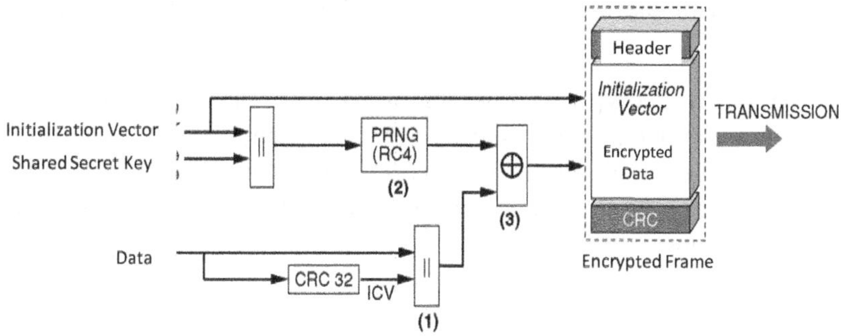

Figure 8.3. *The WEP encryption process*

This method, using a 64- or 128-bit key, is easily broken due to the 24-bit initialization vector, easy to identify, and the key to a length of 40 or 104 bits. To break a 104-bit key, at the most you can now get a million encrypted frames. Breaking the key takes only a few minutes on a standard server. Methods for forcing an AP need to recover a million packets, often much less. This can be performed in a few minutes. It is strongly recommended this protection solution is not used. Two authentication techniques are associated with WEP:

– open system authentication;

– shared key authentication.

In the first, which is the default authentication system, authentication is explicit. A terminal can therefore be associated with any AP and listen to all the data that travel within the BSS. The second is better because it uses a shared secret key mechanism. The shared key authentication mechanism involves four steps:

1) a station wishing to associate with an AP sends an authentication frame;

2) when the AP receives this frame, it sends a frame composed of 128 bits to the station, i.e. a random text generated by the WEP algorithm;

3) after receiving the frame containing the text, the station encrypts the text with the shared secret key before sending this new text to the AP;

4) the AP decrypts the encrypted text using the same shared secret key and compares it with the one that was sent earlier. If the text is identical, the AP confirms the correct authentication, otherwise the AP sends a negative authentication frame. Figure 8.4 shows these different steps.

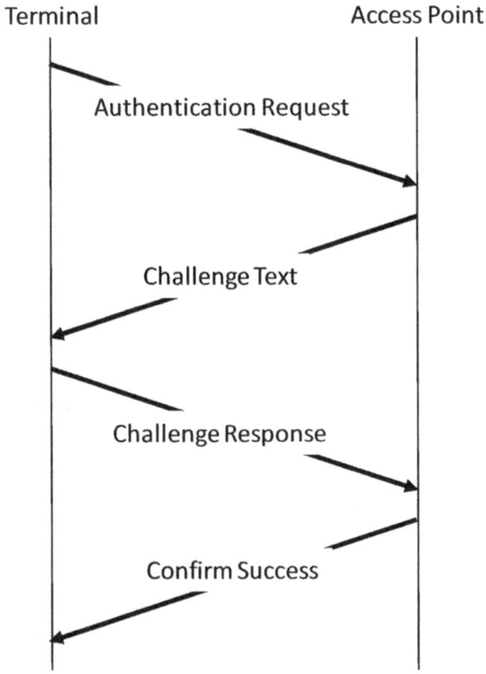

Figure 8.4. *Operation of the shared key authentication mechanism*

As the key used for this authentication is the same as for the encryption process, from the moment the key has been broken, authentication is also broken. A significant improvement of these mechanisms was made by the introduction of the IEEE 802.1x protocol. We describe this scheme in detail a little later in this chapter.

8.2.8. *WPA and IEEE 802.11i*

We highlighted the weaknesses of WEP. In June 2004 the IEEE 802.11i Working Group finalized an architecture designed to fill these gaps WPA2. Between the time when WEP was deprecated and the arrival of WPA2, the WiFi Alliance supported the interim WPA recommendation. Based on a subset of 802.11i, WPA uses the existing hardware (AP and wireless card). WPA2 is not compatible with the first generation of WiFi equipment. The main difference between WPA and WPA2 encryption is the protocol used: RC4 and AES for WPA and WPA2, respectively.

The contributions of the IEEE 802.11i standard can be classified into three categories:

– definition of multiple wireless security protocols;

– information leading to the choice of one protocol;

– new method of key distribution.

The 802.1x standard for authentication uses the calculation of a master key, called the pairwise master key. In ad-hoc mode, this key is called the pre-shared key and distributed manually. A robust security network (RSN) 802.11i, or secure network, then uses 802.1x for authentication and key management. The access control is based on a strong authentication of the upper layers. The RSN must guarantee security and mobility, integrity and confidentiality, as well as scalability and flexibility.

8.3. Security and authentication

The security architecture must provide client authentication, regardless of whether home or foreign network. An architecture with centralized authentication server – Remote Authentication Dial-In User Server (RADIUS) – can satisfy this requirement. The customer no longer has to worry about the AP to which it is associated.

In 802.1x architecture, network communication has three elements: a "supplicant", which is the 802.1x station asking to be authenticated, an "authenticator", the AP, and an authentication server, usually RADIUS.

Each authenticator shares a secret with the RADIUS server with which it communicates. This secret is used to compute an HMAC-MD5 summary of RADIUS packets exchanged. Each packet contains a RADIUS Request Authenticator field, which is an HMAC-MD5 digest of the package, calculated with this secret. This field is inserted by the RADIUS server and verified by the AP. In the other communication direction, the RADIUS server checks the EAP Authenticator attribute with the EAP message attribute. These two attributes offer a mutual authentication packet and preserve the integrity of the communication between the RADIUS server and the AP.

Since it is relatively easy for an attacker with suitable picking tool to listen to traffic between the stations on the wireless link, the proposed security architecture is designed to provide strong guarantees of confidentiality. It further defines a dynamic key distribution mechanism.

8.3.1. *Scalability and flexibility*

The 802.11i architecture is scalable to the number of mobile users supported. A user moves from one AP to another and may be reauthenticated quickly and securely. To meet the deployment needs of wireless networks in businesses and public places, this security architecture is flexible in order to facilitate administration and to take into account the existing deployment environment.

By separating the authenticator of the authentication process itself, the RSN enables scaling of the number of APs. The flexibility is provided by the fact that the optional Extensible Authentication Protocol (EAPoL)-Key message may be disabled for a particular deployment, where the confidentiality of data is not necessary.

The model specifies how the 802.11i RSN interacts with 802.1x. Two types of protocols provide security at the MAC level:

– Temporal Key Integrity Protocol (TKIP);

– Counter with Cipher Block Chaining Message Authentication Code Protocol.

A transitional security network supports pre-RSN architecture, especially the following mechanisms inherited from 802.11:

– Open Authentication;

– Shared Key Authentication;

– WEP.

An RSN network must support CCMP. NRT can, however, make the transition with previous networks by implementing TKIP. Figure 8.5 illustrates the different levels of security that must be supported for a global security architecture.

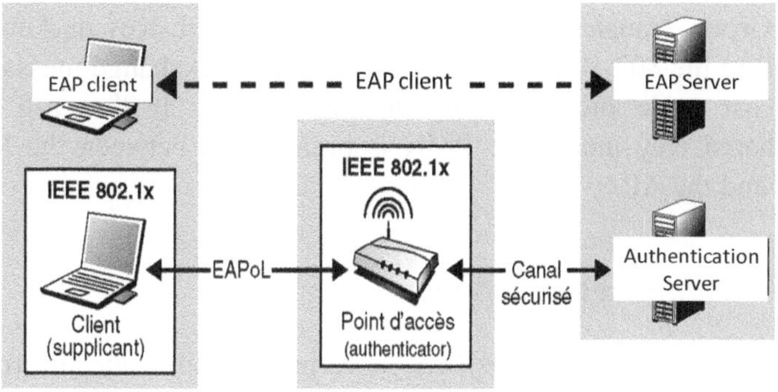

Figure 8.5. *Security levels in the 802.1x architecture*

The authenticator (the AP) and the authentication server perform mutual authentication and establish a secure channel.

8.3.2. *IEEE 802.11i*

The model does not describe the methods used to carry out this operation; protocols such as RADIUS, IPSec or Transport Lager Security/Secure Socket Layer can be implemented.

The 802.1x client or supplicant and the authentication server authenticate each other using the Extensible Authentication Protocol and generate a master key PMK. The elements of this procedure are carried by the secure channel. The cryptographic parameters are different for each 802.1x client. The Pairwise Master Key (PMK) is shared between the 802.1x client and the AP. They use a four-pass protocol, or 4-way handshake, and EAPoL-Key messages for the following operations:

1) confirmation of the existence of the PMK;

2) confirmation of the commissioning of the PMK;

3) calculation of key Pairwise Transient Key (PTK) from the PMK;

4) setting up encryption keys and integrity of 802.11 frames;

5) confirmation of the setting function of 802.11 keys.

This process is shown in Figure 8.6.

A Group Transient Key, transmitted via EAPoL-Key packets from the AP to the 802.1x client, allows the latter to exchange messages in broadcast mode and optionally in unicast mode. In the case of the Pre-Shared Key mode, the PMK is preinstalled between the 802.1x client and the AP.

8.3.3. *Trading security policy*

An AP broadcasts in its beacon or probe frames the information element to notify the 802.1x client of:

– list of authentication infrastructure supported (typically 802.1x);

– list of security protocols available (TKIP, CCMP, etc.);

– encryption method for the distribution of a group key (GTK).

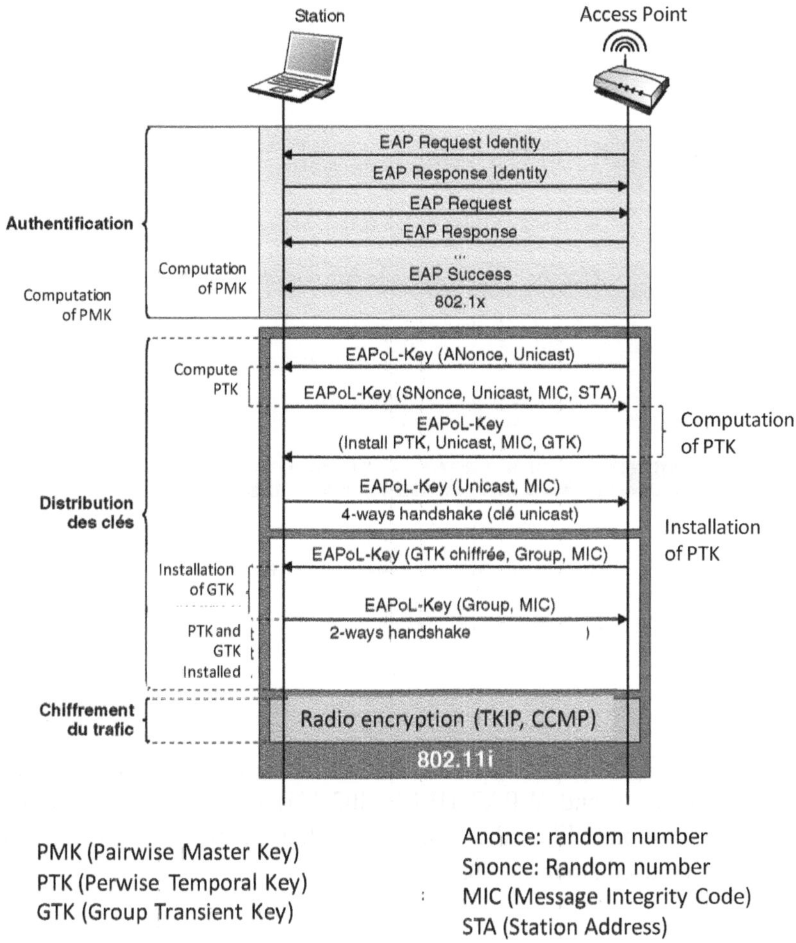

Figure 8.6. *Operation of the 802.11i protocol for passes*

An 802.11 this choice is notified via an information element inserted in its request for association. This approach is illustrated in Figure 8.7.

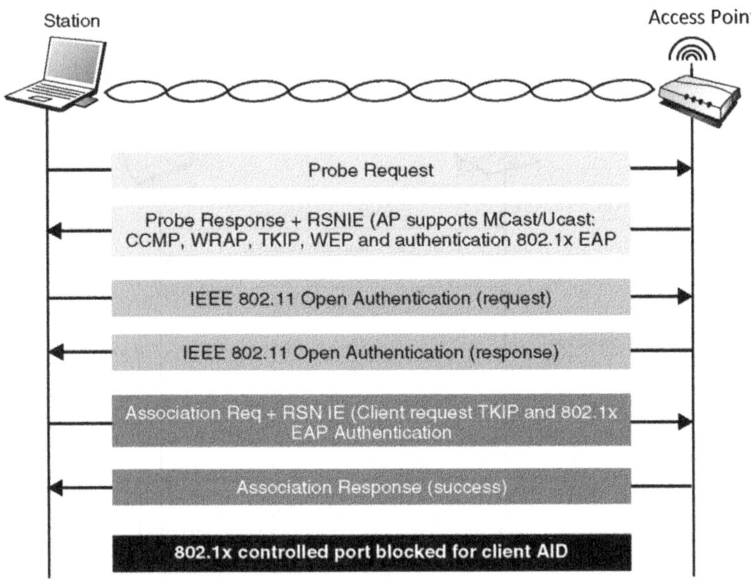

Probe Request : requête d'estampille
RSN IE (Robust Security Network Information Element)
CCMP (Counter-mode/CBC-MAC Protocol)
TKIP (Temporal Key Integrity Protocol)
WEP (Wired Equivalent Privacy)

Figure 8.7. *Negotiating security policy*

The security of WiFi networks started inauspiciously. Today, even outside of WPA and WPA2 (IEEE 802.11i) there are many highly reliable solutions like the use of virtual private networks or technology-based smart cards.

8.4. Saving energy

Wireless terminals can be fixed or mobile. The main problem of mobile terminals is their battery life, which generally has little autonomy. To increase the terminal activity time, the IEEE 802.11

standard provides a power-saving mode. Specifically, there are two modes of terminal operation:

– continuous aware mode;

– power save polling mode.

The first is the default operation in which the station is working all the time and constantly listens to the media. The second allows energy savings. In this case, the AP maintains a record of all the stations that are in energy-saving mode and stores the data addressed to them. Stations that are on standby are activated at regular times to receive a particular frame, the Traffic Information Map (TIM) frame, sent by the AP.

TIM frames are transmitted between AP and terminals. All stations share the same time interval for receiving of TIM frames. TIM frames allow mobile devices to know whether or not they have some data stored in the AP. When a terminal activates and receives a TIM frame, this allows us to determine that the AP contains data. The terminal sends the AP a request, called a Polling Request Frame, to initialize the data transfer. Once the transfer is complete, it returns to standby until reception of the next TIM frame.

For broadcast or multicast traffic type, the AP sends a Delivery Traffic Information Map frame to the terminal.

WiFi network IP packets must be transmitted over the wireless medium. For this, they must be placed in an Ethernet frame. To control and manage the connection, it is necessary to have some specific frames. The three types of frames available in WiFi are:

– data frame, for the transmission of user data;

– control frame, to control access to the media (RTS, CTS and ACK);

– management frame, to associate a station with an AP, as well as for synchronization and authentication.

All wireless frames are composed in the manner shown in Figure 8.8.

CRC (Cyclic Redundancy Cheksum) PLCP (Physical Convergence Protocol) MAC (Medium Access Control)

Figure 8.8. *Structure of a wireless frame*

The preamble is dependent on the physical layer and contains the following two sequences:

– Synch of 80 bits alternating between 0 and 1 which is used by the physical circuit to select the antenna;

– Start Frame Delimiter, a sequence of 16 bits, 0000 1100 1011 1101, used to define the beginning of the frame: FXP246.

The Physical Layer Convergence Protocol (PLCP) header contains the following logical information used by the physical layer to decode the frame:

– word length PLCP_PDU: represents the number of bytes in the packet, which allows the physical layer to correctly detect the end of the packet;

– PLCP signaling: contains information about the transmission speed between the adapter card and the AP;

– error control header field: error detection CRC field of 16 bits.

The MAC transports the underlying protocol, as shown in Figure 8.9.

8.5. IEEE 802.11a, b and g

Basic WiFi networks come from the IEEE 802.11b standard on the 2.4 GHz band. As explained above, this standard originated from studies in the general 802.11 Working Group framework.

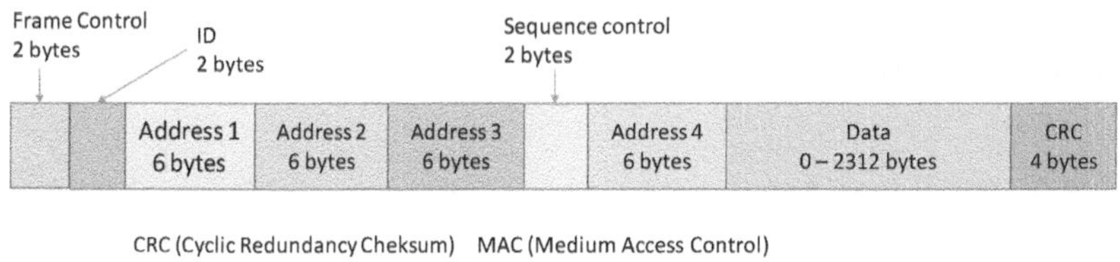

Figure 8.9. *The MAC zone*

8.5.1. *IEEE 802.11b*

This proposal has become a standard, and billions of WiFi cards have been sold. First deployed on campuses, airports, train stations and large public and private administrations, this solution allows the connection of laptops, PDAs or smartphones in corporate networks. IEEE 802.11a and g proposals increase the capacity to more than 50 Mbps. IEEE 802.11n was a new extension on both 2.4 and 5.15 GHz, allowing several hundreds of Mbps.

IEEE 802.11 networks work with APs whose peak rates are 11, 54 and 600 Mbps and have a range of a few tens of meters. For this maximum carrier value, it is necessary for the terminal to be close enough to the AP, less than a block away. We need to engineer the network, while calculating the positioning of the various APs.

Thirteen frequencies are available in the United States in the 2,400–2,483.5 MHz band, so a total width of 83.5 MHz, and 14 in Europe. Three carriers maximum can be used because the transmission requires a bandwidth of 20 MHz which runs on 25 MHz. These frequencies, graded, from 1 to 13, make it possible to use 1, 6 and 11 frequencies as the carrier.

The 802.11b and 802.11g networks are compatible in the upward direction. An 802.11g card can connect to an 802.11b network at a speed of 11 Mbps, but the reverse is impossible. In contrast, the 802.11b/g and 802.11a frequencies being totally different, there is no compatibility between the two solutions. If the device that wants access to both networks has two access cards, the frequency can, however, overlap.

8.5.2. *IEEE 802.11a*

The IEEE 802.11a was the first working group but only the second proposed after IEEE 802.11b. For the physical part, the following proposals were selected for 802.11a:

– frequency in the 5GHz frequency band U-NII without license (Unlicensed-National Information Infrastructure), which requires no license;

– OFDM modulation with 52 carriers, allowing excellent performance. A total of eight independent channels are available;

– eight rates, between 6 and 54 Mbps, are available. The rate selected by the access card depends on the reception strength.

The maximum distance between the card and the AP may exceed 100 m, but the fall in communication rate is strongly linked to distance. For the rate of 54 Mbps, the mobile station containing the WiFi card can be a few meters away from the AP. The flow drops very quickly to be approximately equivalent to that which would be achieved with 802.11b at 100 m.

By using small cells, so that the frequencies are highly reusable, 802.11a allows tens of customers per 100 m^2. Such a network is therefore able to support high-quality video streaming.

Layers above the MAC layer match those encountered in Ethernet networks.

8.5.3. *IEEE 802.11n*

With IEEE 802.11n pre-products, the maximum speed reached was 108 Mbps, in late 2007. This is double IEEE 802.11a and 802.11g standards. IEEE 802.11n networks offer a potential peak rate of

600 Mbps. In fact, the actual flow rate is much lower and is in the order of 100 Mbps in the best case.

The goal of the IEEE 802.11n standard is threefold:

– to make changes to MAC and physcial levels so that the peak rate reaches 600 Mbps. For this, a MIMO technology is implemented;

– to greatly improve useful throughput at the application level to get 100 megabits per second;

– to stay compliant with IEEE 802.11a and 802.11g.

The first and last goals are clearly understandable. The second is not so clear but we can recall the disastrous actual flow rates of the first WiFi networks: 5–6 Mbps at best in the case of 802.11b and some 20 Mbps for 802.11a and 802.11g. The objective of the new standard is to provide an effective transfer rate of about 100 Mbps.

IEEE 802.11n is a milestone in product compatibility, since it includes the IEEE 802.11i standard for security and confidentiality of data transport over the air interface. This standard is not compatible with the first two generations, especially with the use of Advanced Encryption Standard (AES). The increase in speed is an opportunity to encourage users to completely change their wireless network infrastructure.

The IEEE 802.11e standard for offering a quality of service around DiffServ type technology is included in 802.11n. This integration allows us to prioritize the different flows that pass through the wireless network.

IEEE 802.11n allows native handover management by integrating the IEEE standard 802.11f. On a wireless network, it is possible to switch from one AP to another AP, but the handover occurs only at a low speed comparable to the speed of a pedestrian.

MIMO technology is not new, but it hit the market about 2010 due to a very complex implementation. MIMO aims to carry multiple streams in parallel on different antennas but using the same frequency.

The idea is to connect multiple antennas at the transmitter, which emit different flows on the same frequency. Due to the properties of multipath found in environments with obstacles, the signals arrive at different times at the receiver. If the receiver is powerful enough, it is able to decrypt the binary sequences arriving at different times. Note that the MIMO technique is satisfactory in environments with obstacles and use outside buildings greatly reduces its flow.

The MIMO operation diagram is shown in Figure 8.10, assuming n transmitters and n receivers. MIMO can operate with only one receiving antenna, but with the risk of not recovering all signals. In this figure, the same bit sequence is transmitted over all antennas. The objective is to increase the quality of the transmission since the receiver gets the same bits several times and chooses the most probable value. An odd number of antennas have to be chosen to be sure to separate the probabilities of the bits received. To increase the quality of the reception by transmitting the same bit several times, diversity means increasing the throughput by transmitting different bits. This second solution is illustrated in Figure 8.11.

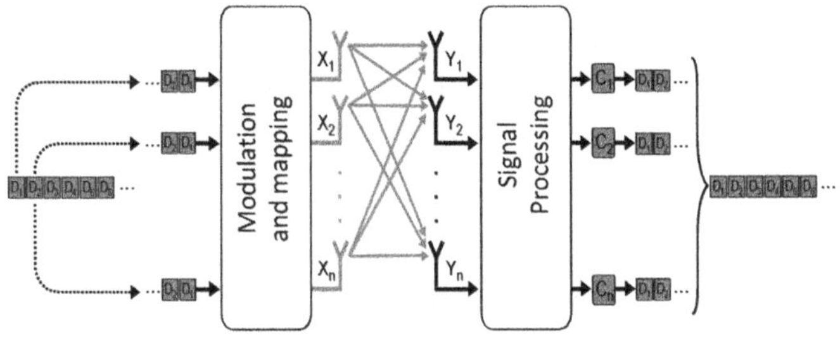

Figure 8.10. *MIMO technique with increased quality*

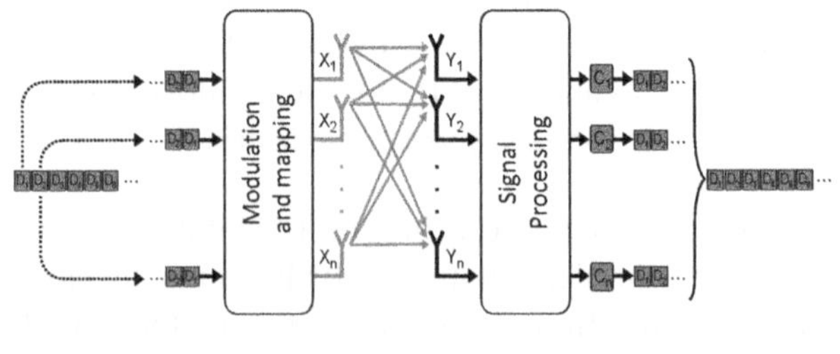

Figure 8.11. *MIMO technique with increased capacity*

8.5.4. *IEEE 802.11ac*

IEEE 802.11ac is a version of WiFi designed to surpass 1 Gbps, using multiple antennas, which can run simultaneously but in separate sectors. Four mutually complementary solutions have been put in place to obtain the target data rates. The first, which is fairly simple, consists simply of increasing the capacity of the transmission channel using the 5 GHz band, which is much freer than the 2.4 GHz band. The range of this higher frequency band is a little less. In addition, obstacles hamper the signals somewhat. However, the available bandwidth is 200 MHz, which is much greater than the 83.5 MHz assigned to the 2.4 GHz band. This enables us to have WiFi channels whose frequency is 80 MHz, and optionally, 160 MHz. These different channels are shown in Figure 8.12.

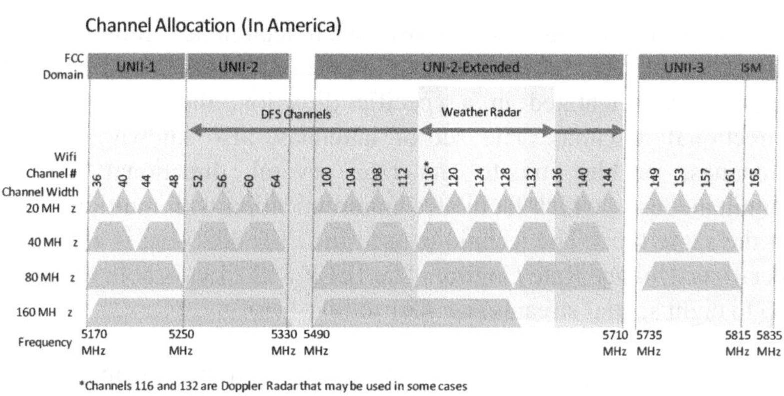

Figure 8.12. *The 5 Ghz band*

The second solution pertains to directionality of the signals emitted, using a set of specific antennas. This technique involves allowing the transmission of several communications on the same frequency, but in different directions. There is spatial multiplexing, which gives the name of this technique: Space Division Multiple Access (SDMA). Figure 8.13 illustrates this technology.

Figure 8.13. *SDMA*

SDMA technology uses transmission techniques allowing us to direct the signals – better known as "beamforming" – whereby the signals are orientated in a specific direction, thus giving rise to directional antennas. The set of antennas, also known as virtual antennas, enables us to generate several directional antennas simultaneously, which facilitates communication with multiple clients at the same time. The technique used in IEEE 802.11ac is also called Per-User Unitary Rate Control. The IEEE 802.11ac is able to support up to eight spatial streams (vs. four in 802.11n).

The third 802.11ac advance comes from multiple antennas able to deliver MU-MIMO (Multi-User MIMO) – that is to connect several pairs of clients to one another, using several antennas simultaneously on the same frequency, while minimizing interference. In other words, each directional antenna can be replaced by several antennas with the same directionality, using the same frequency to carry out communications in MIMO mode.

The fourth solution to increase the capacity of the network is modulation: 256-QAM, rate 3/4 and 5/6, added as optional modes (vs. 64-QAM, rate 5/6 maximum in 802.11n). Even, some vendors offer a non-standard 1024-QAM mode, providing 25% higher data rate compared to 256-QAM.

With this technology, we are able to multiply the data rate by the number of directional antennas. In the example shown in Figure 8.13, the two physical antennas each give rise to three virtual antennas, which is a total of six virtual antennas. These six antennas mean that six simultaneous communications can be established. With these six antennas, we can also establish three 2 × 2 MIMO communications (two virtual antennas communicating with two antennas on a remote machine). We can just as well have three simultaneous communications: one 3 × 3 MIMO communication, one 2 × 2 MIMO communication and a one-way communication.

The peak data rate reaches 250 Mbps per virtual antenna. With four physical antennas each bearing three virtual antennas, we obtain 12 virtual antennas. Thus, the total data rate is 3 Gbps. Also, if the bands used are wider – say, 80 MHz instead of 20 MHz – then theoretically we can achieve over 10 Gbps. In actual fact, as for all WiFi APs, the peak data rate is only achieved in exceptional cases, where there is no external interference and the clients are located very near to the antenna. In reality, the nominal data rates are often divided by at least a factor of 2, but often much more, to obtain the real data rates.

The group IEEE 802.11ac, with 433 Mbps per antenna over an 80 MHz band and a MIMO with two antennas per client, delivers 866 Mbps per connection with a terminal. With four physical antennas, we achieve over 3Gbps. With a bandwidth of 160 MHz, the data rate is doubled. The standard specifies that there may be up to eight spatial sectors, which again doubles the data rate. Obviously, these reported figures are peak data rates, which are only very rarely achieved in reality. As is the case with all WiFi solutions, fallback rates become necessary when the conditions are not ideal. Compatibility with existing standards also requires the adaptation of data rates to conform to standards which are not compatible with IEEE 802.11ac.

The aim with the new IEEE 802.11ax standard (Super-fast WiFi) is to further increase the data rate by using more numerous and more directional antennas.

8.5.5. *IEEE 802.11ad*

The IEEE 802.ad standard was introduced in Chapter 6 as a WPAN solution under the name of WiGig. But as this is a WiFi solution, we have summarized this solution in this section. IEEE 802.11ad uses frequencies in the range of 60 GHz. This standard is championed by a group of manufacturers: the Wireless Gigabit Alliance. The basic product could be tri-band WiGig, operating simultaneously on the 2.4, 5 and 60 GHz bands, and therefore able to adapt to the environment. The peak data rate is 7 Gbps. The range in the 60 GHz band is less than 10 m, so this network is said to belong to the category of Personal Area Networks.

The channels used in the base mode correspond to 57.24, 59.4, 61.56 and 63.72 GHz. The bandwidth is 2.16 GHz. To compensate for the significant attenuation of the signal, IEEE 802.11ad requires directional antennas which can focus the radio beam within a 6° angle. With this very narrow beam, the IEEE 802.11ad standard is used outdoors between directional antennas with a direct view, or indoors but counting on ricochets to reach the addressee. Indeed, with a 6° beam, the connection between two machines is often complex. Fortunately, there is an option which enables us to direct the beam so that the two machines wishing to connect can communicate with each other.

Two applications of IEEE 802.11ad are mentioned in the standard. The first corresponds to wireless connections for peripheral computers, with the aim of doing away with the numerous cables which trail everywhere. This application also facilitates simple sharing of peripheral devices. The second application relates to consumer electronics, associated with wireless technology, such as stereo equipment, HD televisions and online gaming systems.

8.5.6. *IEEE 802.11af*

The IEEE 802.11af technique pertains to a totally different method to increase the data rate: the use of cognitive radio. This consists of reusing frequency bands which are not being used at a time t in the television bands. More generally, recent measurements show that, in spite of their scarcity and their high price, frequencies between 0 and 20 GHz are underused – quite often less than 10%.

The only bands which are heavily used are those employed by telecom operators which, owing to the TDMA, CDMA and OFDMA techniques, exhibit excellent degrees of utilization. The bands reserved for TV channels, which are below 1 gigahertz, are particularly attractive for operators. These TV channels could be used by APs to achieve ranges of hundreds of meters. Such ranges demonstrate that WiFi technology is capable of serving the needs of WiMAX and 4G clients.

The WiFi standard IEEE 802.11af – which must not be confused with IEEE 802.3af, or Power over Ethernet – is called TV White Space. The expression "white space" refers specifically to those frequencies which are not used by wireless television channels. The operational bandwidth for cognitive radio covers dozens of digital television channels. The data rates achieved by this method may be very high indeed. The name of the product is "White-Fi", also sometimes called "Super WiFi".

One important question relates to the way in which cognitive radio is used. It is for each state to regulate this usage. The main solutions pertain to the way in which unoccupied frequencies are used. A number of solutions suggest themselves, depending on the coverage of the cognitive AP. If the band is not occupied at the AP's location, it can be assumed that it is not occupied within a radius of a few meters of that point. This category includes WiFi APs inside buildings with limited power, only covering the inside of the building. A second solution is to refer to a bandwidth management server, which knows where the bands are used. Thus, a common discipline needs to be used between the primary and secondary users. Standardization work is being carried out in different working groups, such as IEEE P1900, with a view to proposing solutions acceptable for both parties. In the United States, the proposal is to have APs and stations able to determine their position using a satellite positioning system such as GPS and use the Internet to query a geolocation database provided by a regional regulatory agency to discover which frequency channels are available for use at a given time and position. A third, more long time solution could stem from the ability for each device to measure the interference and indicate to the secondary APs whether or not they can exploit the frequency.

At the physical level, the method used in IEEE 802.11af is Fixed Subcarrier Spacing with OFDM. The channels used in OFDM may or may not be contiguous – that is they may belong to television channels that are directly beside one another or that are separated by active television channels.

A channel contains 64 bearers. Four channels may be selected, each operating at 6, 7 or 8 MHz. In the latter two cases, as the data rate per channel reaches 26.7 Mbps and given that, as with IEEE 802.11ac, we can have at least four directions in SDMA, the maximum peak rate is 568.9 Mbps, with 8 MHz channels.

The bearer access technique is the same as with normal WiFi. Service classes using an Enhanced Distributed Channel Access technique from IEEE 802.11e are available, and correspond to the four classes background, best-effort, video and voice. A fifth, even higher, priority is added for spectrum-monitoring to determine which channels could potentially be used.

An important issue concerns the range of the IEEE802.11af solution. To avoid interference, the surface covered by the AP could be small if the emitting power is low, e.g. 1–10 mW. If the power is approaching 100 mW, the range is pretty large. A comparison is given in Figure 8.14.

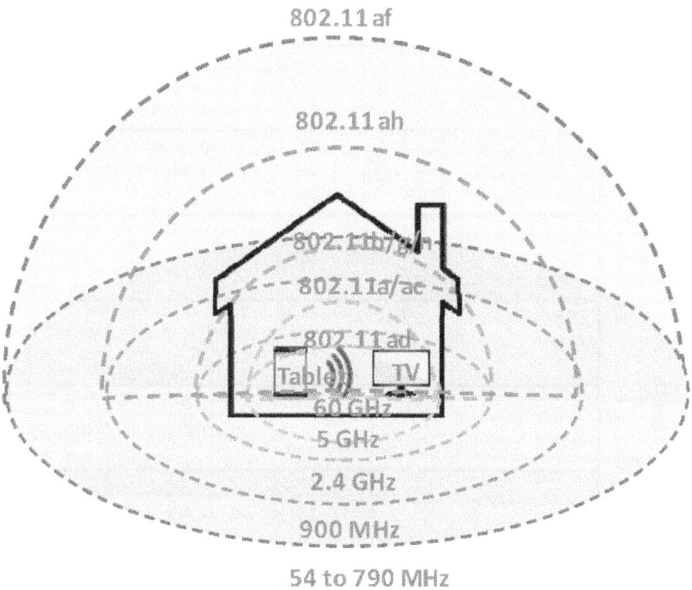

Figure 8.14. *Range comparison of the different WiFi solutions*

8.5.7. *IEEE 802.11ah*

The IEEE is working on a new standard, adapted to the connection of "things" – i.e. sensors or other small devices with low consumption and often minimal cost. This solution works in the non-licensed bands available below 1 GHz. The different possibilities are described in Figure 8.15.

The technique should enable us to connect a large number of things with low consumption. The sharing of the transmissions takes place in classic WiFi mode, but with a much lower data rate and a much greater range, of up to 1 km. The things are grouped into clearly-defined sets so as to avoid too high a number of simultaneous accesses and so as not to lose too much in contention in CSMA/CD mode. In addition, IEEE 802.11h involves the use of classic techniques in the world of Ethernet to put stations on standby and reawaken them.

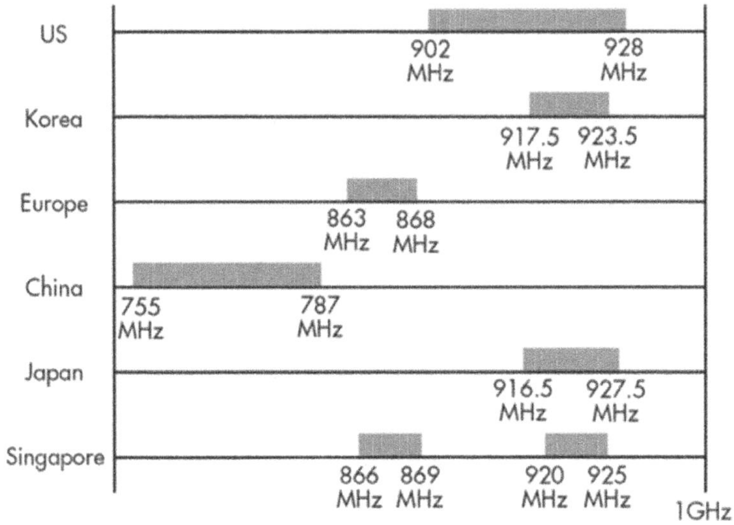

Figure 8.15. *The 900 MHz bands*

IEEE 802.11h introduces the function of relay, whereby certain things can connect through the relay, which decreases competition by limiting the number of objects attempting to transmit simultaneously. Relay increases the geographic scope of the things that can be connected and, at the same time, saves energy by having tasks managed locally by the relay. In the standard, the aim is not to create a mesh network, as the number of hops is limited to two.

Energy savings are also of crucial importance in the current context, where consumption is increasing exponentially, and the IEEE 802.11ah standard is taking steps in this direction. For this purpose, the connected machines are divided into two classes: TIM (Traffic Indication Map) terminals and non-TIM terminals. TIM terminals periodically receive information about the traffic on the AP. Non-TIM terminals use the new mechanism Target Wake Time (TWT) to reduce the overhead of the signaling procedure.

TWT is a function which allows an AP to define a time, or a set of times, for certain terminals to have access to the communication bearer. The terminal and the AP exchange information including the allotted duration of the activity, to enable the AP to control any competition and clashes between competing terminals. The AP can reserve time periods for a terminal to transmit due to various protective mechanisms. The use of TWT is negotiated in advance between an AP and a terminal. The TWT mechanism is also used to reduce energy consumption, because stations using it can enter a semi-sleep state until their TWT comes around.

Another important mechanism in IEEE 802.11ah is the Restricted Access Window, which limits the number of terminals which have the right to use the AP. This limitation is achieved by putting the terminals into groups. Channel access is restricted solely to those terminals belonging to a given group, over the course of a set period. This mechanism helps reduce competition and avoid simultaneous transmissions from too high a number of machines which are unaware of one another's presence.

The IEEE also defines a Transmission Opportunity, wherein an AP and a client can carry out a set of exchanges during a reserved period. The purpose of this mode of operation is to reduce the number of competing channels, improve the efficacy of the channels by minimizing the number of exchanges and help extend the terminals' battery life by minimizing their periods of activity; the rest of the time, the terminal is on standby.

The division of the coverage area of a BSS into sectors, determining subsets of terminal machines, is called "sectorization". It is done by a set of antennas or a set of directional antenna lobes to cover different sectors of the BSS. The goal of sectorization is to decrease contention and interference by reducing the number of terminals that can transmit simultaneously.

8.6. Conclusion

WiFi is doing well. It is always subject to strong expansion and the number of WiFi components sold every day is in the millions. The quality of service was a failure for the first generations of WiFi. This is no longer a problem with overcapacity in the recent versions that exceeds the Gbps and allows each user to connect to tens of Mbps.

8.7. Bibliography

[ALA 05] AL AGHA K., PUJOLLE G., VIVIER G., *Réseaux de mobiles et réseaux sans fil*, 2nd ed., Eyrolles, 2005.

[ALE 01] ALESSO H.P., SMITH C.F., *The Intelligent Wireless Web*, Addison Wesley, 2001.

[CAL 03] CALLAWAY E.H., *Wireless Sensor Networks: Architectures and Protocols*, Auerbach Publications, 2003.

[CHA 06] CHANDRA P., LIDE D., *Wi-Fi Telephony: Challenges and Solutions for Voice over WLANs*, 1st ed., Newnes, 2006.

[DAV 01] DAVIS H., MANSFIELD R., *The Wi-Fi Experience: Everyone's Guide to 802.11b Wireless Networking*, Que, 2001.

[DAV 03] DAVIES J., *Deploying Secure 802.11 Wireless Networks with Microsoft Windows*, Microsoft Press, 2003.

[DOR 02] DORMAN A., *The Essential Guide to Wireless Communications Applications*, Prentice Hall, 2002.

[FLI 01] FLICKENGER R., *Building Wireless Community Networks*, O'Reilly, 2001.

[GAS 05] GAST M.S., *802.11 Wireless Networks: The Definitive Guide*, 2nd ed., O'Reilly, 2005.

[GEI 06] GEIER E., *Wi-Fi Hotspots: Setting Up Public Wireless Internet Access*, Cisco Press, 2006.

[GRA 07] GRAHAM A.W., KIRKMAN N.C., PAUL P.M., *Mobile Radio Network Design in the VHF and UHF Bands: A Practical Approach*, Wiley, 2007.

[HAC 03] HAC A., *Wireless Sensor Network Designs*, Wiley, 2003.

[HEL 03] HELTZEL P., *Complete Wireless Home Networking*, Prentice Hall, 2003.

[JAN 03] JANEVSKI T., *Traffic Analysis and Design of Wireless IP Networks*, Artech House, 2003.

[KHA 03] KHAN J., KHWAJA A., *Building Secure Wireless Networks with 802.11*, Wiley, 2003.

[LAR 02] LA ROCCA J., *802.11 Demystified: Wi-Fi Made Easy*, McGraw-Hill, 2002.

[MAL 03] MALLICK M., *Mobile and Wireless Design Essentials*, Wiley, 2003.

[MAL 04] MALES D., PUJOLLE G., *Wi-Fi par la pratique*, 2nd ed., Eyrolles, 2004.

[MAX 02] MAXIM M., POLLINO D., *Wireless Security*, McGraw-Hill, 2002.

[MUR 04] MURTHY C.S.R., MANOJ B.S., *Ad Hoc Wireless Networks: Architectures and Protocols*, Prentice Hall, 2004.

[NIC 01] NICHOLS R.K., LEKKAS P.C., *Wireless Security: Models, Threats, and Solutions*, McGraw-Hill, 2001.

[OHA 99] O'HARA B., PETRICK A., *The IEEE 802.11 Handbook: a Designer's Companion*, IEEE Press, 1999.

[OHR 03] OHRTMAN F., ROEDER K., *Wi-Fi Handbook: Building 802.11b Wireless Networks*, McGraw-Hill, 2003.

[PAH 01] PAHLAVAN K., KRISHNAMURTHY P., *Principles of Wireless Networks: a Unified Approach*, Prentice Hall, 2001.

[PER 00] PERKINS C., *Ad-hoc Networking*, Addison Wesley, 2000.

[PRA 98] PRASAD R., *Universal Wireless Personal Communications*, Artech House, 1998.

[RAP 01] RAPPAPORT T.S., *Wireless Communications Principles and Practice*, Prentice Hall, 2001.

[ROS 03] ROSHAN P., LEARY J., *Wireless Local-Area Network Fundamentals*, Cisco Press, 2003.

[SAN 94] SANTAMARIA A., LOPEZ-HERNANDEZ F., *Wireless LAN Systems*, Artech House, 1994.

[SAY 01] SAYRE C.W., *Complete Wireless Design*, McGraw-Hill, 2001.

[STA 01] STALLINGS W., *Wireless Communications & Networks*, Prentice Hall, 2001.

[SWE 04] SWEENEY D., *WiMax Operator's Manual: Building 802.16 Wireless Networks*, Apress, 2004.

[TOH 01] TOH C.K., *Ad-hoc Mobile Wireless Networks: Protocols and Systems*, Prentice Hall, 2001.

[WAN 03] WANG X., POOR H.V., *Wireless Communication Systems: Advanced Techniques for Signal Reception*, Prentice Hall, 2003.

[WEB 00] WEBB W., *Introduction to Wireless Local Loop*, Artech House, 2000.

[ZHA 04] ZHAO F., GUIBAS L., *Wireless Sensor Networks*, Morgan Kaufmann Publishers, 2004.

[ZHA 06] ZHANG Y., HU H., FUJISE M., *Resource, Mobility, and Security Management in Wireless Networks and Mobile Communications*, 1st ed., Auerbach, 2006.

9

WMAN and WiMAX

IEEE 802.16 is a series of wireless broadband standards written by the Institute of Electrical and Electronics Engineers (IEEE). The IEEE Standards Board established a working group in 1999 to develop standards for broadband for wireless metropolitan area networks. The Workgroup is a unit of the IEEE 802 local area network and metropolitan area network standards committee.

Although the 802.16 family of standards is officially called WirelessMAN in IEEE, it has been commercialized under the name "WiMAX" (from "Worldwide Interoperability for Microwave Access") by the WiMAX Forum industry alliance. The Forum promotes and certifies compatibility and interoperability of products based on the IEEE 802.16 standards. The 802.16e-2005 amendment version was announced as being deployed around the world in 2009. This version is called Mobile WiMAX.

The Mobile WiMAX based on the IEEE 802.16e standard (the mobile version of the IEEE 802.16-2004 standard) was expected to shift from fixed subscribers to mobile subscribers with various form factors: Personal Digital Assistant (PDA), phone or laptop. Hence, it was expected that the Mobile WiMAX would not only compete with the broadband wireless market share in urban areas with DSL, cable and optical fibers, but also threaten the hotspot-based WiFi and even the voice-oriented cellular wireless market. This is due to the variety of fundamentally different design options. For example, there are multiple physical layer (PHY) choices: a single-carrier-based physical layer called Wireless-MAN-SCa, an orthogonal frequency division multiplexing (OFDM)-based physical layer called Wireless MAN-OFDM and an orthogonal frequency division multiple access (OFDMA)-based physical layer called Wireless-OFDMA. Similarly, there are multiple choices for medium access control (MAC) architecture, duplexing, frequency band of operation, etc.

However, for practical reasons of interoperability, the scope of the standard needs to be reduced, and a smaller set of design choices for implementation need to be defined. The WiMAX Forum does this by defining a limited number of system and certification profiles. Accordingly, WiMAX Forum has defined mobility system profiles for IEEE 802.16e, which we use in the entire dissertation for the design of the wireless communication system and the performance evaluation of the Mobile WiMAX in the link and the system levels.

In this chapter, we present a concise technical overview of the emerging Mobile WiMAX solution for broadband wireless. The purpose here is to provide an executive summary for the most important issues related to the Quality of Service (QoS) in Mobile WiMAX, mobility, PHY and MAC layers.

9.1. Background on IEEE 802.16e

Mobile WiMAX based on IEEE 802.16e is a wireless broadband solution that offers a rich set of features with a lot of flexibility in terms of deployment options and potential service offering. These features provided by both MAC and PHY layers deserve highlighting as follows:

9.1.1. *The medium access control layer*

The MAC layer of Mobile WiMAX provides a medium-independent interface to the PHY layer and is designed to support the wireless PHY layer by focusing on efficient radio resource management. The MAC layer supports both Point-to-Multipoint (PMP) and Mesh network modes and is divided into three sublayers:

Sublayers: the service-specific convergence sublayer, common part sublayer and security sublayer. The primary task of service-specific convergence sublayer is to classify external Service Data Units (SDUs) and associate each of them with a proper MAC Service Flow (SF) identifier and connection identifier. The common part sublayer function is to (1) segment or concatenate the SDUs received from higher layers into the MAC Protocol Data Units (PDU), (2) retransmit MAC PDUs that were received erroneously by the receiver when Automated Repeat Request (ARQ) is used, (3) provide QoS control and priority handling of MAC PDUs belonging to different data and signaling bearers and (4) schedule MAC PDUs over the PHY resources. The security sublayer handles authentication, secure key exchange and encryption. Among all these functions, we will put the accent on QoS-related functions and mechanisms that are associated with our field of interest and our proposed solutions, as described in the next section.

9.1.2. *Channel access mechanism*

In mobile WiMAX, the MAC layer at the Base Station (BS) is fully responsible for allocating bandwidth to all Mobile Stations (MSs), in both uplink and downlink. It supports several mechanisms by which an MS can request and obtain uplink bandwidth. Depending on the particular QoS and traffic parameters associated with a service, one or more of these mechanisms may be used by the MS. The BS allocates dedicated or shared resources periodically to each MS, which it can use to request bandwidth. This process is called polling. Mobile WiMAX defines a contention access and resolution mechanism for the case when more than one MS attempts to use the shared resource. If it already has an allocation for sending traffic, the MS is not polled. Instead, it is allowed to request more bandwidth by (1) transmitting a stand-alone bandwidth request or (2) piggybacking a bandwidth request on generic MAC packets.

9.1.3. *Quality of service*

Support for QoS is a fundamental part of the mobile WiMAX MAC layer design, strong QoS control is achieved by using a connection-oriented MAC architecture, where all downlink and uplink connections are controlled by the serving BS. Before any data transmission happens, the BS and the MS establish a unidirectional logical link, called a connection, between the two MAC-layer peers. Each connection is identified by a Connection Identifier (CID), which serves as a temporary address for data transmission over the particular link.

Mobile WiMAX also defines a concept of a service flow. An SF is a unidirectional flow of packets with a particular set of QoS parameters and is defined by a Service Flow Identifier (SFID). To support a variety of applications, mobile WiMAX defines five SFs:

– unsolicited grant services (UGS): this is designed to support fixed-size data packets at a Constant Bit Rate (CBR). Examples of applications that may use this service are T1/E1 emulation and VoIP without silence suppression. The SF parameters that define this service are maximum sustained traffic rate, maximum latency, tolerated jitter and request/ transmission policy;

– real-time polling services (rtPS): this service is designed to support real-time SFs such as MPEG video that generate variable-size data packets on a periodic basis. The mandatory SF parameters that define this service are minimum reserved traffic rate, maximum sustained traffic rate, maximum latency and request/transmission policy;

– extended real-time variable rate (ertPS) service: this service is designed to support real-time applications, such as VoIP with silence suppression, which have variable data rates but require guaranteed data rate and delay. The mandatory SF parameters that define this service are minimum reserved traffic rate, maximum sustained traffic rate, maximum latency and request/transmission policy;

– non real-time polling service (nrtPS): this service is designed to support delay-tolerant data streams, such as an FTP, which require variable-size data grants at a minimum guaranteed rate. The mandatory SF parameters to define this service are minimum reserved traffic rate, maximum sustained traffic rate, traffic priority and request/transmission policy;

– best-effort (BE) service: this service is designed to support data streams, such as Web browsing, which do not require a minimum service-level guarantee. The mandatory SF parameters to define this service are maximum sustained traffic rate, traffic priority and request/transmission policy.

9.1.4. *Mobility support*

In addition to fixed broadband access, mobile WiMAX envisions four mobility-related usage scenarios: nomadic, portable, simple mobility up to 60 kmph and full mobility up to 120 kmph. The IEEE 802.16e defines a framework for supporting mobility management.

9.2. The physical layer

The mobile WiMAX physical layer is based on OFDMA. OFDMA is the transmission scheme of choice to enable high-speed data, video and multimedia communications and is used by a variety of commercial broadband systems, including DSL, WiFi, etc. In this section, we cover the basics of OFDMA and provide an overview of the Mobile WiMAX physical layer.

9.2.1. *Subchannelization in Mobile WiMAX: OFDMA*

OFDM is a very powerful transmission technique. It is based on the idea of dividing a given high-bit-rate data stream into several parallel lower bit-rate streams and modulating each stream on separate carriers often called subcarriers. OFDM is a spectrally efficient version of multicarrier modulation, where the subcarriers are selected such that they are all orthogonal to one another over the symbol duration, thereby avoiding the need to have non-overlapping subcarrier channels to eliminate intercarrier interference.

In order to have multiple user transmissions, a multiple access scheme such as Time Division Multiple Access (TDMA) or Frequency Division Multiple Access (FDMA) has to be associated with OFDM. In fact, an OFDM signal can be made from many user signals, giving the OFDMA multiple access. The multiple access has a new dimension with OFDMA. A downlink or an uplink user will have a time and a subcarrier allocation for each of its communications. However, the available subcarriers may be divided into several groups of subcarriers called subchannels. Subchannels may be constituted using either contiguous subcarriers or subcarriers pseudo randomly distributed across the frequency spectrum. Subchannels formed using distributed subcarriers provide more frequency diversity. This permutation can be represented by Partial Usage of Subcarriers (PUSC) and Full Usage of Subcarriers (FUSC) modes.

The subchannelization scheme based on contiguous subcarriers in mobile WiMAX is called Band Adaptive Modulation and Coding (AMC). Although frequency diversity is lost, band AMC allows system designers to exploit multiuser diversity, allocating subchannels to users based on their frequency response.

9.2.2. *Slot and frame structure in OFDMA-based mobile WiMAX*

Before giving details of mobile WiMAX frame structure, it is worth mentioning that the downlink and uplink transmissions coexist according to one of the two duplexing modes: Time Division Duplex (TDD) or Frequency Division Duplex (FDD). They are sent through the downlink and uplink subframes. However, TDD is favored by a majority of implementations of WiMAX forum. The frame is divided into two subframes: a downlink subframe followed by an uplink subframe after a small guard interval. Figure 9.1 shows an OFDMA frame when operating in TDD mode. The downlink subframe begins with a downlink preamble that is used for physical layer procedures, such as time and frequency synchronization and initial channel estimation. The download preamble is followed by a Frame Control Header (FCH), which provides frame configuration information, such as the MAP message length, the modulation and coding scheme, and the usable subcarriers. Multiple users are allocated data regions within the frame, and these allocations are specified in the Uplink and downlink MAP messages (DL-MAP and UL-MAP) that are broadcasted following FCH in the downlink subframe. MAP messages include the burst profile for each MS, which defines the modulation and coding scheme used in that link.

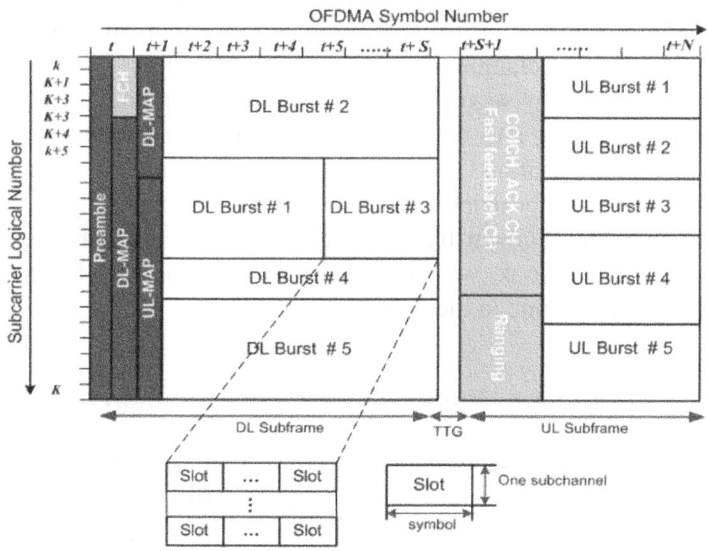

Figure 9.1. *A simple mobile WiMAX OFDMA frame structure for the TDD mode*

The uplink subframe is made up of several uplink bursts from different users. A portion of the uplink subframe is set aside for contention-based access that is used for a variety of purposes. This subframe is used mainly as a ranging channel to perform closed-loop frequency, time and power adjustments during network entry as well as periodically afterward. The ranging channel may also be used by MS to make uplink bandwidth requests. Besides the ranging channel and traffic bursts, the uplink subframe has a Channel Quality Indicator Channel (CQICH) for the MS to feedback channel-quality information that can be used by the BS scheduler and an Acknowledgment (ACK) channel for the MS to feedback downlink acknowledgments.

According to what we described in the previous section, once higher layer data have been classified into SFs and scheduled by the MAC layer, they are assigned to OFDMA slots by a slot allocator. A slot is the basic resource unit in OFDMA frame structure as it is a unit of (subchannel-symbol). One can consider that the data region (OFDMA frame) is a two-dimensional allocation which can be visualized as a rectangle. Allocating OFDMA slots to data in the downlink is done by segmenting the data after the modulation process into blocks that fit into one OFDMA slot. It is considerable to note that the definition of an OFDMA slot depends mainly on the mode of permutation of subcarriers in an OFDMA subchannel, i.e. FUSC, PUSC and AMC. Given that we focus on AMC mode for resource allocation in this book, we will detail only the structure of slots in an AMC OFDMA frame in the following section.

9.2.3. *OFDMA slot structure in AMC permutation mode*

Mobile WiMAX supports a variety of AMC schemes and allows for the scheme to change on a burst-by-burst basis per link, depending on channel conditions. In the downlink, Quadrature Phase Shift Keying (QPSK), 16 Quadrature Amplitude Modulation (QAM) and 64 QAM are mandatory for mobile WiMAX; 64 QAM is optional in the uplink. Forward Error Correction (FEC) using convolutional codes is mandatory. Convolutional codes are combined with an outer Reed-Solomon code in the downlink for OFDM-PHY. The standard optionally supports turbo codes and Low Density Parity Check (LDPC) codes at a variety of code rates as well. A total of 52 combinations of modulation and coding schemes are defined in mobile WiMAX as burst profiles. In order to constitute an OFDMA frame in the AMC, it is important to understand the slot structure that is called band AMC mode.

Unique to the band AMC permutation mode, all subcarriers constituting a sub channel are adjacent to each other. Therefore, having a microscopic view of an OFDMA frame in AMC mode, nine adjacent subcarriers with eight data subcarriers and one pilot subcarrier are used to form a bin, as shown in Figure 9.2. Four adjacent bins in the frequency domain constitute a band. An AMC slot consists of six contiguous bins within the same band. Thus, an AMC slot can consist of one bin over six consecutive symbols, two consecutive bins over three consecutive symbols or three consecutive bins over two consecutive symbols. Therefore, in each frame, MSs are allocated a successive set of slots, forming bursts. Each allocation is represented in the DL-MAP message by the slot offset and the number of slots in the allocation frame.

Generally, the BS receives periodically the Channel Quality Indicator feedback from the MSs indicating information such as channel-quality measurements such as Received Signal Strength Indicator and Signal-to-interference-plus-noise ratio (SINR). Then, the BS supporting AMC allows the scheduler to exploit multiuser diversity by allocating each MS to its corresponding strongest sub channel. Hence, increasing the overall throughput of the system.

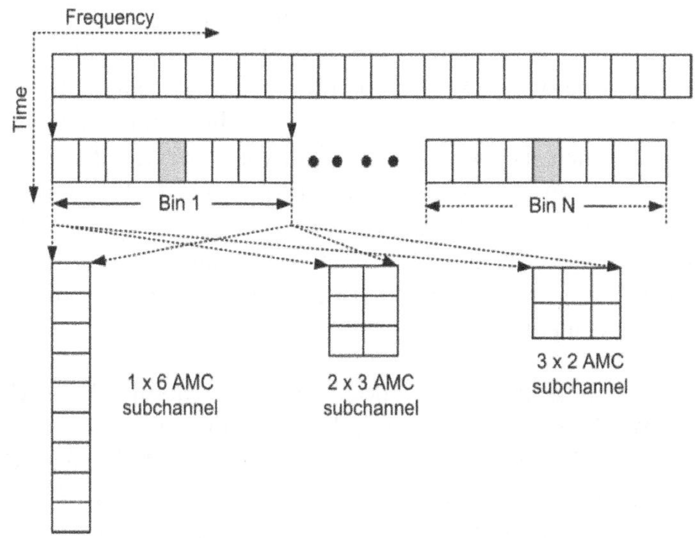

Figure 9.2. *Mapping OFDMA slots to subchannels and symbols in IEEE*

9.3. An example of WiMAX and WiFi integration

The general WiFi/Mobile WiMAX interworking environment we consider is illustrated in Figure 9.3. We adopt the interworking architecture based on loose coupling. The necessary changes in both WiFi and Mobile WiMAX systems are rather limited as it will integrate both systems at the IP layer and relies on the IP protocol to handle mobility between access networks.

The main characteristic of this architecture is to assume two overlapped cells of a Mobile WiMAX and a WiFi, where both cells are served by a BS and an access point respectively. We assume that the AP is connected to the WiFi access network which can have a dedicated gateway to the Mobile WiMAX and traffic from Mobile WiMAX to WiFi or vice versa will be routed through this gateway. The MN has dual interfaces: WiMAX and WiFi.

As shown in Figure 9.3, the Mobile WiMAX supports access to variety of IP multimedia services via WiMAX radio access technologies which is called Access Service network (ASN). The ASN is owned by a Network Access Provider and comprises one or more BSs and one or more ASN gateways (ASN-GW) that form the radio access network. Access control and traffic routing for MSs in Mobile WiMAX are entirely handled by the Connectivity Service Network, which is owned by a Network Service Provider, and provides IP connectivity and all the IP core network functions. The WiFi access network may be owned either by the NAP or by any other part (e.g. public WiFi operator or an airport authority), in which case the interworking is enabled and governed by appropriate business and roaming agreement.

For enabling the interworking of WiFi/Mobile WiMAX, the Mobile WiMAX CSN core network incorporates four new functional elements: the Mobile WiMAX AAA server, the CSN-GW, the WiFi access gateway (WAG) and the Packet Data Gateway (PDG). The WiFi must also support similar interworking functionality to meet the access control and routing enforcement requirements. The Mobile

WiMAX AAA server in the Mobile WiMAX domain terminates all AAA signaling originated in the WiFi that pertains to HMS. This signaling is typically based on radius or diameter. The Mobile WiMAX AAA server interfaces with other Mobile WiMAX components, such as the WAG and PDG. The Mobile WiMAX AAA server can also route AAA signaling to/from another Mobile WiMAX AAA server, in which case it serves as a proxy and is referred to as Mobile WiMAX AAA proxy. To support mobility, the Foreign Agents (FA) located in the ASN Gateway are considered as the local FAs in the interworking architecture. However, for enabling vertical handover, a Mobile IP Home Agent (HA) has been added to the architecture. While the HA may be local to either network, it must be accessible by both networks.

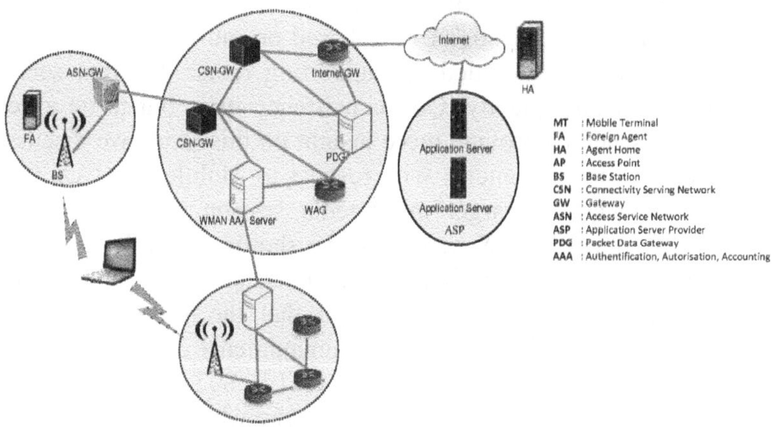

Figure 9.3. *WiFi/WiMAX interworking architecture*

As shown in Figure 9.3, traffic from HMS is routed to the WAG and finally to the PDG. This routing is enforced by establishing appropriate traffic tunnels after a successful access control procedure. The PDG function is much like a CSN-GW in Mobile WiMAX domain. It routes the user data traffic between the HMS and an external packet data network (in our case, the IP multimedia network), and serves as an anchor point that hides the mobility of the HMSs within the WiFi domain. The WAG functions mainly as a route policy element, ensuring that user data traffic from authorized HMS is routed to the appropriate PDGs, located in either the same or a foreign Mobile WiMAX.

9.3.1. *QoS management*

One vital component for the provision of seamless multimedia session continuity is QoS consistency across the WiFi and Mobile WiMAX networks. This is indeed vital because without QoS consistency the multimedia sessions will experience different QoS levels in both network domains, and hence seamless continuity will not be achievable. It is unfortunate, however, that the WiFi and Mobile WiMAX specifications were based on rather different sets of requirements, and they ended up supporting rather different sets of QoS features. Consequently, QoS consistency turns out to be a very challenging issue. To provide more insight on this issue, we discuss a list of WiFi QoS deficiencies with respect to Mobile WiMAX QoS. Apparently, when we target multimedia session continuity across WiFi and Mobile WiMAX networks, we should carefully take these deficiencies into consideration and understand their impacts. The discussion is based on the assumption that the WiFi MAC layer complies with IEEE 802.11 plus the amendments of IEEE 802.11e, and the physical layer complies with IEEE 802.11g. The mobile version of WiMAX is based on IEEE 802.16e while the fixed version on IEEE 802.16d.

9.3.2. *Qos support and classes*

IEEE 802.11 standard is intended to support only best effort service, however, IEEE 802.11e introduced basic QoS support by defining four different access categories (ACs), namely AC_VO (voice) with highest priority, AC_VI (video), AC_BE (best effort) and AC_BK (background) with lowest priority.

User priority	*Category(AC)*	*Informative*
1	AC_BK	Background
2	AC_BK	Background
0	AC_BE	Best Effort
3	AC_BE	Video
4	AC_VI	Video
5	AC_VI	Video
6	AC_VO	Voice
7	AC_VO	Voice

Table 9.1. *IEEE 802.11e access category and user priority mapping*

While in IEEE 802.16e, the QoS is represented by five SFs: Unsolicited Grant Service (UGS) for (VoIP without silence suppression), Real-time polling service (rtPS) for (video), Extended Real-time polling service (ertPS) for (voice with silence suppression), Non Real-time polling service (nrtPS) for (FTP) and best effort. From the difference in the supported QoS classes, it becomes clear that a vertical handover from Mobile WiMAX to WiFi needs to involve a QoS mapping procedure. Therefore, we have to define for each WiFi QoS class or AC, which Mobile WiMAX SF type we assign and additionally we have to provide a mapping from Traffic Specification negotiated in WiFi in Action. ADDT requests to Dynamic Service Addition Request (DSA-REQ) negotiated in Mobile WiMAX.

9.4. Mechanisms of channel access

An important aspect to consider is that the basic support for QoS differs significantly between WiFi and Mobile WiMAX due to their different PHY and MAC layers design. While the access to the channel in Mobile WiMAX is completely centralized, it can be distributed or centralized in WiFi based 802.11e. This is explained in the next section.

9.4.1. *WiFi access methods*

It is worth describing the mandatory access method in the legacy 802.11 in order to compare it with the 802.11e. The access method is based on Distributed Coordination Function (DCF). The basic DCF uses a Carrier Sense Multiple Access with Collision Avoidance (CSMA/CA) mechanism to regulate access to the shared wireless medium. Before initiating a transmission, each WS is required to sense the medium and perform a binary exponential backoff. If the medium has been sensed as being idle for a time interval called DCF Interframe Space (DIFS), the WS enters a backoff procedure. A slotted backoff time is generated randomly from a Contention Window (CW) size: backoff time = rand [0, CW] x slot time. In DCF, only best effort service is provided. Time-bounded multimedia applications (e.g. voice over IP and videoconferencing) require certain bandwidth, delay and jitter guarantees. The point is that with DCF, all the WSs compete for the channel with the same priority. There is no differentiation mechanism to provide better service for real-time multimedia traffic than for data applications. This is the reason behind introducing the hybrid coordination function in the IEEE 802.11e which consists of two different methods of medium access and uses the concepts of Traffic Opportunity (TXOP), which refers to a time duration during which a WS is allowed to transmit a burst of data frames:

1) The Enhanced Distributed Channel Access (EDCA) method in which each AC behaves as a single DCF contending entity with its own contention parameters (CWmin, CWmax, AIFS and TXOP), which are announced by the AP periodically in beacon frames. Basically, the smaller the values of CWmin, CWmax and AIFS[AC], the shorter the channel access delay for the corresponding AC and the higher the priority for access to the medium. In EDCA, a new type of IFS is introduced, the Arbitrary IFS (AIFS), instead of DIFS in DCF. Each AIFS is an IFS interval with arbitrary length as follows: AIFS = SIFS + AIFSN × {slot time}, where AIFSN is called the arbitration IFS number. After sensing that the medium has been idle for a time interval of AIFS[AC], each AC calculates its own random backoff time (CWmin[AC] ≤ backoff time ≤CWmax[AC]). The purpose of using different contention parameters for different queues is to give a low-priority class a longer waiting time than a high-priority class, so

the high-priority class is likely to access the medium earlier than the low-priority class.

2) The polling-based HCF Controlled Channel Access (HCCA) method in which different traffic classes called traffic streams (TSs) are introduced. Before any data transmission, a traffic stream (TS) is first established, and each WS is allowed to have no more than eight TSs with different priorities. In order to initiate a TS connection, a WS sends a QoS request frame containing a Traffic Specification (TSPEC) to the AP. A TSPEC describes the QoS requirements of TS, such as mean/peak data rate, mean/maximum frame size, delay bound and maximum Required Service Interval (RSI). On receiving all these QoS requests, the AP scheduler computes the corresponding HCCA-TXOP values for different WSs by using their QoS requests in TSPECs (TXOP1, TXOP2, etc.) and polls them sequentially.

9.4.2. *Mobile WiMAX access method*

In Mobile WiMAX, the MAC layer at the base station is fully responsible for allocating bandwidth to all users, in both the uplink and the downlink. The only time the MS has some control over bandwidth allocation is when it has multiple sessions or connections with the BS. Depending on the particular QoS and traffic parameters associated with a service, one or more of these mechanisms may be used by the MS. The BS allocates dedicated or shared resources periodically to each MS, which it can use to request bandwidth. This process is called polling. Polling may be done either individually (unicast) or in groups (multicast). Multicast polling is done when there is insufficient bandwidth to poll each MS individually. When polling is done in multicast, the allocated slot for making bandwidth requests is a shared slot which every polled MS attempts to use. Mobile WiMAX defines a contention access and resolution mechanism for the case when more than one MS attempts to use the shared slot. If it already has an allocation for sending traffic, the MS is not polled. Instead, it is allowed to request more bandwidth by (1) transmitting a stand-alone bandwidth request MPDU, (2) sending a bandwidth request using the ranging channel or (3) piggybacking a bandwidth request on generic MAC packets.

9.4.3. *Handover support*

Handover in WiFi is almost always hard, which means break-then-make approach. This is because the WiFi station cannot be serviced in parallel by more than one AP and therefore has to break its communication with its current AP before establishing connection with a new one. Hence, considerable transmission disruptions may exist that result in QoS degradation. Moreover, handovers in IEEE 802.11 are solely controlled by the WiFi AP, so the WiFi infrastructure cannot provide tight control of the QoS provisioning.

However, in IEEE 802.16e, there are three types of handover: (1) hard handover, (2) fast base station switching (FBSS) and (3) macro diversity handover (MDHO). In the last two methods, the MS maintains a valid connection simultaneously with more than one BS. In the FBSS case, the MS maintains a list of the BSs involved, called the active set. The MS continuously monitors the active set, does ranging and maintains a valid connection ID with each of them. The MS, however, communicates with only one BS, called the anchor BS. When a change of anchor BS is required, the connection is switched from one base station to another without having to explicitly perform handover signaling. The MS simply reports the selected anchor BS on the CQICH. Macro diversity handover is similar to FBSS, except that the MS communicates on the downlink and the uplink with all the base stations in the active set-called a diversity set here- simultaneously. In the downlink, multiple copies received at the MS are combined using any of the well-known diversity-combining techniques. In the uplink, where the MS sends data to multiple base stations, selection diversity is performed to pick the best uplink.

9.5. IEEE 802.16m or mesh for WiMAX

WiMAX networks can be structured into two operating modes: Point-to-Multipoint and mesh networks. Mesh mode is a type of operation that can interconnect multiple mobile clients together with many WiMAX base stations (nodes) and form a network of connections so as to provide a wide coverage area for mobile clients. All the clients can communicate with each other and there is no need for an intermediate node to act as the mediator of the network. In this mode, the IEEE 802.16 can provide broadband access with wireless support both single-hop and multi-hop settings. The basic topology of an IEEE 802.16 mesh network consists of two participating entities, called BS and subscriber station (SS), displayed below (Figure 9.4). The BS is the central node, responsible for coordinating all the communication and providing connectivity to the client stations (fixed or mobile).

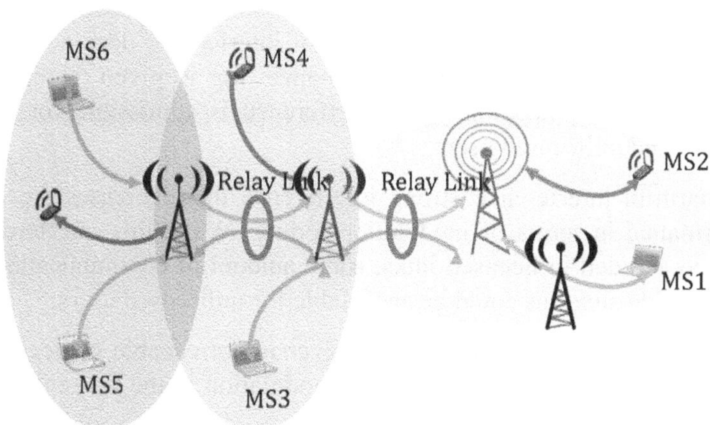

Figure 9.4. *Topology of an IEEE 802.16 mesh network*

9.6. IEEE 802.16h or cognitive radio for WiMAX

The final 802.16h IEEE standard defines a set of cognitive radio capabilities for WiMAX networks.

The proposed global framework is divided into two separate profiles. The first one provides uncoordinated coexistence mechanisms (WirelessMAN-UCP), i.e. without requiring much interaction among the different systems and hence adequate for heterogeneous systems, while the second provides coordinated coexistence mechanisms (WirelessMAN-CX), which addresses the required coordination of neighboring systems in order to reduce the interference generated to each other.

9.6.1. *Uncoordinated coexistence mechanism*

In a real scenario, a WiMAX network may coexist with licensed users (denoted in the amendment as specific spectrum users) and other unlicensed users (denoted as non-specific spectrum users) sharing the same frequency band. In such a case, the interference that the WiMAX system may cause to each of the users can be different. The standard IEEE 802.16h defines three possible levels of interference:

– acceptable interference: this level of interference does not cause degradation in the receiver performance for a given choice of modulation and/or coding. This interference is admissible for both licensed and unlicensed users;

– harmful interference: strong interference that decreases the link performance in terms of modulation/coding. While this interference must be avoided in licensed links, some amount of communication is still possible and thus could be acceptable for unlicensed users;

– destructive interference: the receiver is not capable of decoding the received signal for any available modulation at the transmitter. It must be avoided.

In order to achieve these acceptable interference levels, both to licensed users and unlicensed users, the standard provides a set of mechanisms. These include:

– testing channels for other users;

– discontinuing operations after detecting channel activity;

– detecting other users;

– scheduling for channel testing;

– requesting and reporting measurements by different nodes;

– selecting and advertising a new channel.

While all these procedures are described in detail in the standard, the timing and threshold parameters used are left open and must be specified by each regulatory administration. In the case of unlicensed users, the proposed procedures are still valid with minor variations. For example, instead of performing a search for channels free of licensed users, the cognitive network looks for the best set of channels for operation either when certain unlicensed users are present.

The standard 802.16h, in the uncoordinated profile, permits distributed architectures for the radio resource management within the network formed by one 802.16 BS and its associated subordinated nodes. Each BS has a Distributed Radio Resource Management entity to execute the spectrum sharing policies of 802.16h and to build up a database for sharing information related to actual and intended future usage of radio spectrum. This database can be recovered from a master entity with the required information or from different devices (e.g. using the GPS, IP address, operator information, radio signature scheduling info, etc.).

In order to avoid the regulation infringement, the different interferers must be identified by their radio signature, which can be a short preamble, peak power, relative spectral density, etc. Every transmitter will send the radio signature during an interference-free slot. The time position of this slot (frame number, sub-frame and time-shift) will be used for identification.

Once the environmental data are available, the Radio Resource Management entity performs a real-time, adaptive scheduling, which can be done in terms of channel or even interference free regions within a MAC frame.

9.6.2. *Coordinated coexistence mechanism*

When multiple secondary networks coexist in the same region, they can collaborate in order to coordinate their transmissions and build a neighbor relationship. In the standard 802.16h, the following three basic mechanisms for achieving coexistence are provided:

– MAC Frame Synchronization, including Tx and Rx intervals, for separating transmissions and enabling operation in synchronized zones;

– Dynamic Channel Selection and Adaptive Channel Selection for finding a less interfered or less used frequency (similar to the uncoordinated case);

– separation of the remaining interference in the time domain, by using a Coexistence Frame, coordinated scheduling and a fairness approach, thus allowing the usage of a frequency channel by more than one system.

9.7. Bibliography

9.7.1. *Standards*

IEEE 802.11E/D6.0 EEE Draft Standard Medium Access Control (MAC) Quality of Service (QoS) Enhancements, November 2003.

IEEE 802.16E, Standard for Local and Metropolitan Area Networks – Part 16: Air Interface for Fixed and Mobile Broadband Wireless Access Systems, February 2006.

IEEE Standard for Local and metropolitan area networks Part 16 – Air Interface for Broadband Wireless Access Systems Amendment 2: Improved Coexistence Mechanisms for License-Exempt Operation.IEEE Std 802.16h-2010 (Amendment to IEEE Std 802.16-2009). July 30 2010. doi: 10.1109/IEEESTD.2010.5538195.

IEEE P802.21/D02.00 IEEE, Standard for Local and Metropolitan Area Networks, Media Independent Handover Services, 2006.

9.7.2. *Selected bibliography*

[AHS 07] AHSON S.A., ILYAS M., *WiMAX Handbook*, CRC, 2007.

[AND 07a] ANDREWS J.G., GHOSH A., MUHAMED R., *Fundamentals of WiMAX Understanding Broadband Wireless Networking*, Pearson Education, Inc., 2007.

[AND 07b] ANDREWS J.G., GHOSH A., MUHAMED R., *Fundamentals of WiMAX: Understanding Broadband Wireless Networking*, 1st ed., Prentice Hall, 2007.

[EKL 06] EKLUND C., MARKS R.B. , PONNUSWAMY S. *et al.*, *Wireless MAN: Inside the IEEE 802.16 Standard for Wireless Metropolitan Area Networks*, IEEE Standards Information Network/IEEE Press, 2006.

[HAR 06] HARTE L., WIBLITZHOUSER A., PAZDERKA T., *Introduction to 802.16 WiMax, Wireless Broadband Technology, Operation and Services*, Althos Publishing, 2006.

[HAR 07] HARTE L., KALAICHELVAN K. , *WiMax Explained; System Fundamentals*, Althos, 2007.

[LEE 08] LEE B.G., CHOI S., *Broadband Wireless Access & Local Networks: Mobile Wimax and Wifi*, 1st ed., Artech House Publishers, 2008.

[NUA 07] NUAYMI L., *WiMAX: Technology for Broadband Wireless Access*, Wiley, 2007.

[OHR 05] OHRTMAN F., *WiMAX Handbook*, 1st ed., McGraw-Hill Professional, 2005.

[OHR 06] OHRTMAN F., *WiMAX in 50 Pages*, Monnoz Publishing, 2006.

[OLE 04] OLEXA R., *Implementing 802.11, 802.16, and 802.20 Wireless Networks: Planning, Troubleshooting, and Operations*, Newnes, 2004.

[SAL 04] SALKINTZIS A.K., "Interworking techniques and architectures for WLAN/3G integration toward 4G mobile data networks", *IEEE Wireless Communication*, vol. 11, no. 3, pp. 50–61, February 2004.

[SHE 06] SHEPARD S., *WiMAX Crash Course*, 1st ed., McGraw-Hill Osborne Media, 2006.

[SWE 05] SWEENEY D., *WiMax Operator's Manual: Building 802.16 Wireless Networks*, 2nd ed., Apress, 2005.

[THI 07] Third Generation Partnership Project, Technical Specification Group Services and System Aspects, Architecture enhancements for non-3GPP accesses (Release 8), 3GPP TS 23.402, 2007.

[UNG 08] UNGER J., *Deploying License-Free Wireless Wide-Area Networks*, 1st ed., Cisco Press, 2008.

[WAK 07] WAKIKAWA R., IPv4 Support for Proxy Mobile IPv6 draft-ietf-netlmm-pmip6-ipv4-support-00.txt. IETF 2007.

[WED 99] WEDLUN E., SCHULZRINNE H., "Mobility support using SIP", *Proceedings of ACM International Workshop on Wireless Mobile Multimedia*, vol. 1, pp. 719–737, 1999.

[ZHA 07] ZHANG Y., CHEN H.-H., *Mobile WiMAX: Toward Broadband Wireless Metropolitan Area Networks*, Auerbach, 2007.

10

WRAN and Interconnection

In this chapter, we introduce two different sections. The first one is devoted to IEEE 802.11 which uses Regional Area Network (RAN) technology. RAN technology provides a very large cell with a range of several tens of kilometers. The second section considers the link between the different wireless solutions: how to interconnect Personal Area Networks (PANs), Local Area Networks (LANs), Metropolitan Area Networks (MANs) and RANs.

10.1. IEEE 802.22

A new initiative supported by many manufacturers in the field of regional area networks (RANs) concerns Wireless Regional Area Network (WRAN). This standard, whose study started in 2005, was finalized in 2014. It aims to use the white TV band, that is to say the unused TV channels. The selected frequencies being between 54 and 862 MHz, the range can reach several tens of kilometers and allow applications to broadcast interactive television, interactive video games and video on demand on laptops on the move.

IEEE P802.22.1 is a standard to enhance harmful interference protection for low-power licensed devices operating in TV bands. IEEE P802.22.2 is a recommended practice for the installation and deployment of IEEE 802.22 systems. IEEE 802.2.3 defines a standard for spectrum characterization and occupancy sensing. The three

802.22 task groups (TG1, TG2, and TG3) are writing 802.22.1, 802.22.2, and 82.22.3, respectively.

Currently, some more subgroups have been defined within the IEEE802.22.2 to help the deployment of the solution:

IEEE 802.22.2a – This amendment defines a new clause for Management and Control Plane Interfaces and Procedures to the existing IEEE 802.22-2011 for operation in Very High Frequency (VHF)/Ultra High Frequency (UHF) TV broadcast bands between 54 and 862 MHz. The Management Information Base (MIB) structure enhancements include changes to comply with the ASN.1 format and support for the new clause. Modifications to the existing clause on primitives are intended for Cognitive Radio Capabilities to align it with the content in the MIB clause.

IEEE 802.22b – This amendment specifies alternate Physical Layer and necessary Medium Access Control Layer (MAC) enhancements to IEEE802.22-2011 for operation in VHF/UHF TV broadcast bands between 54 and 862 MHz to support enhanced broadband services and monitoring applications. The standard supports aggregate data rates greater than the maximum data rate supported by the IEEE 802.22-2011. This standard defines new classes of 802.22 devices to address these applications and supports more than 512 devices in a network. This standard also specifies techniques to enhance communications among the devices and makes necessary amendments to the cognitive, security and parameters and connection management clauses. This amendment supports mechanisms to enable coexistence with other 802 systems in the same band.

The WRAN underlying techniques are very innovative with the use of cognitive radio on a long range, to determine, for a given terminal, the direction of the emission, power and frequency use. For this, a central database must be able to answer a query to the antenna indicating the frequency to be used. This solution enables us to increase the spectral use.

Each base station would be armed with a GPS receiver to be able to check the position of the antenna. This information is sent to a centralized server (in the Unites States, this central server is managed by the FCC), which responds with the information about available free TV channels and guard bands in the area of the base station. Other solutions are available as local spectrum sensing, where the base station can decide by itself which channels are available for the communications. A combination of these two approaches is also a possibility to enhance the use of cognitive radio in the TV band.

A third solution is to allow the mobile stations to themselves perform a cognitive sensing. The mobile stations sense the spectrum and send periodic reports to the base station to inform them of the available channels. The base station, with the information gathered, evaluates whether a change is necessary in the channel used, or on the contrary, if it should stay transmitting and receiving in the same one.

The physical layer uses the OFDMA technique on a 6–8 MHz channel corresponding to a TV channel. The maximum global throughput for a range of 30 km is 20 Mbps. Several TV channels can be used simultaneously to increase the total capacity of the system.

Concerning the OFDMA MAC technique used in WRAN system, the initialization and network entry need to accommodate the elements of the spectrum usage flexibility. As there is not a fixed channel for the system, and no pilot channel can be broadcast, any terminal station when turning on and initializing needs to be able to find the signals. Accordingly, when initializing, any mobile station first scans the available spectrum to look at channel occupancy. First, the mobile station has to detect the free channel within the TV band. In the empty channels, the mobile station scans for base station pilot signals and acquires any network information available. Once the mobile station has acquired the correct network, it can then proceed to connect to the network.

Within the OFDM solution, it is necessary to define a format for data transmission. To enable the data to be suitably structured, the transmission is formatted into frames and superframes. The frames are split into two elements: the upstream subframe and the downstream subframe. The boundary between these two subframes is variable and can be adapted to accommodate changes on the levels of upstream and downstream capacity. The superframes are built up from the smaller frames. The superframe is used to provide overall synchronization for the system, and in particular provides the initial network access initialization. At the beginning of each superframe, there is a preamble known as the superframe Control Header. The superframe Control Header contains the information needed for any new mobile stations that need to access the base station.

Considering that an antenna can expect to have 200 MHz of global bandwidth below 1 GHz, which is to say for waves penetrating into buildings, such as television bands, it is obvious that this new regional technology will compete with already established operator solutions. A single antenna with a 200 MHz available bandwidth allows a 20–30 km range that should allow us to group 3,000 interactive television streams or the equivalent of one hundred thousand mobile phones.

10.2. Interconnection between IEEE standards

Future wireless networks aim to provide a universal ubiquitous coverage across different radio technologies through a multi-modal Mobile Node (MN), while offering a rich range of services with variable bandwidth and Quality of Service (QoS) anytime and anywhere. These features require connectivity across multiple networks with different radio technologies, over different geographic areas, with access to different types of services. Such connectivity can be provided by the 4G architecture which envisions highly flexible and adaptive integration of diverse radio technologies to support built-in capabilities for seamless interaction in these environments.

Nevertheless, the integration of heterogeneous networks poses many challenges such as interface selection, seamless handover mechanisms, coordinated configuration of QoS mechanisms and content adaptation strategies. Any mobile system in such integrated networks has to be able to adapt to changing access networks and changing environments automatically without bothering humans to intervene for making mobile computing feasible. To adapt to such changes, the system needs information about the environment, including various pieces of information about the access network itself, information about services supported by this network, as well as more general information about the current context in which the mobile entity resides. Thus, a framework for gathering and processing information is needed.

The evolving IEEE 802.21 standard defines a framework intended to improve the handover decisions of mobile devices based on mainly lower-layer information from both mobile devices and the access network. Such information is provided by defining an abstraction layer called Media Independent Handover (MIH). MIH is used for handover optimization, particularly between heterogeneous networks.

The increased need for intelligence in such environments is behind our motivation to introduce autonomic networking. This is because future networks are expected to embrace the heterogeneity arising from the different network control technologies such that it appears homogenous to the potential users. Therefore, next-generation networks are expected to be self-managed in order to treat automatically network composition and reconfiguration. In this chapter, we address the main issues that arise while implementing the interworking mechanisms between different radio access networks. We provide an interworking mechanism based on the IEEE 802.21 framework, which will be described in detail.

10.2.1. *IEEE 802.21 framework*

The design of the IEEE 802.21 framework is intended to enable seamless handover and interoperability between heterogeneous network types including both 802 and non-802 networks. This is done by introducing a new layer specified by MIHF which provides three main functionalities: media independent events service (MIES), media independent command service (MICS) and media independent information service (MIIS).

10.2.2. *IEEE 802.21 core architecture*

The heart of the IEEE 802.21 framework is the Media Independent Handover Functions (MIHFs), which provide abstracted services to higher layers by means of a unified interface. This unified interface exposes service primitives that are independent of the access technology, and called Service Access Point (SAP). Figure 10.1 illustrates an example showing how the MIHF communicates with access-specific lower layer MAC and PHY components, including 802.16, 802.11 and cellular networks, using lower layer interfaces, and with upper layer entities. The services provided by MIHF are described as follows:

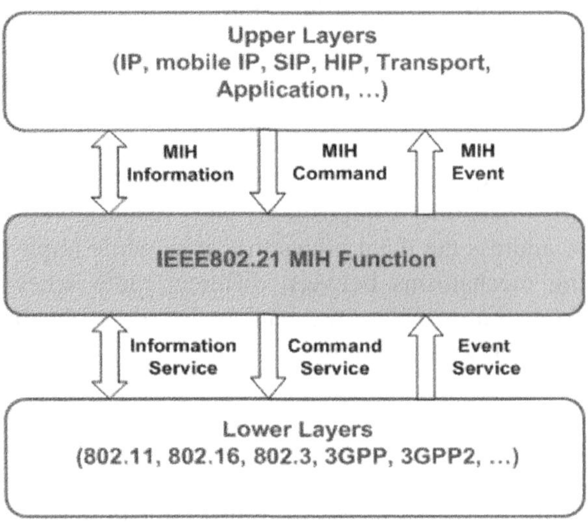

Figure 10.1. *IEEE 802.21 framework*

– Media independent event service: the event service is used to facilitate handover detection. Events inform the condition of the present network and transmission behavior of the data links, radio resource management, etc. The defined events include pre-trigger (L2 Handover Imminent), Link Available, Link Up, Link Parameter Change, Link Going Up, Link Down, Link Going Down, etc.

– Media independent command service: higher layers use the MICS primitives to control the functions of the lower layers. MICS is used to gather information about the status of connected links, as well as to execute higher layer mobility and connectivity decisions on the lower layers. The MIH command can be both local and remote. These include commands from the upper layers to MIH and from MIH to the lower layers.

– Media independent information service: as a mobile node is about to move out of its current network, it needs to discover the available neighboring networks and communicate with the elements within these networks so as to optimize the handover. MIIS provides a framework and corresponding mechanisms by which an MIHF entity can discover and obtain network information within a geographic area. MIIS primarily provides a set of information elements, the information structure and its representation as well as query/response type mechanism. The information service provides access to both static information and dynamic information.

The MIH can be considered as a middleware at the link layer level in the components of network that control the mobility and the mobile device. Such a middleware provides functions to higher layer level. MIH extends the MIP since it can treat the information of layer 2, especially information concerning the status of the network and the use of the multiple technological interfaces. For doing so, the standard uses the triggers in multiple layers and proposes a 2.5 sublayer with an optimization of handover for the mobility in heterogeneous interconnected networks.

IEEE 802.21 also provides the intelligence in the link layer and all information concerning the network to the higher layers in order to optimize the handover among the heterogeneous medium. The standard can support the handover for stationary and mobile users. For the mobile user, the handover can occur when the wireless condition channel changes. While handover for stationary users occurs when users prefer to use another type of network which has, for example, lower load, better QoS, etc.

IEEE 802.21 supports a cooperative usage of mobile users and also the network infrastructure. The mobile user is capable of detecting the available network, and the infrastructure may store information demanded by that network, such as the list of cells in the neighborhood and the localization of mobile equipment. In general, the equipment of the client and the points of attachment (Access point in WiFi or Base Station in WiMAX) may support a multimodal interface and in some case, it can support more than one interface simultaneously.

10.3. Bibliography

10.3.1. *Standards*

IEEE STD P802.21/D02.00, Standard for Local and Metropolitan Area Networks: Media Independent Handover Services, September 2006.

10.3.2. *Selected bibliography*

[AKA 05] AKYLIDIZ I.F., MOHANTY S., XIE J., "A ubiquitous mobile communication architecture for next generation heterogeneous wireless system", *IEEE Communications Magazine*, vol. 43, no. 6, pp. S29–S36, June 2005.

[BUR 02] BURKHART J., *Pervasive Computing: Technology and Architecture of Mobile Internet Applications*, Addison Wesley, 2002.

[CAC 06] CACCACE F., VOLLERO L., "Managing mobility and adaptation in upcoming 802.21- enabled device", *Proceedings of International Conference on Mobile Computing and Networking*, vol. 1, pp. 1–10, November 2006.

[COR 07] CORREIA L.M., Mobile *Broadband Multimedia Networks: Techniques, Models and Tools for 4G,* 1st ed., Academic Press, 2007.

[MOR 06] MORISIO M., TORCHIANO M., *Developing Services for the Wireless Internet*, 1st ed., Springer, 2006.

[MUL 95] MULLER N.J., TYKE L.L., *Wireless Data Networking*, Artech House, 1995.

11

Internet of Things

The Internet of Things started with the idea of connecting to the Internet wireless and wired sensors found in the home, in the office and everywhere in daily life. Then, a major contribution was generated by Radio Frequency Identification (RFID). The Internet of Things aims to connect all objects that can transmit and receive information. These objects relate to terminal equipment as well as smart dusts, RFIDs or sensors. The concept is simple, but there are many problems because the "things" are generally not sophisticated enough to manage and control communications, and to take care of the associated applications. Figure 11.1 describes some different things.

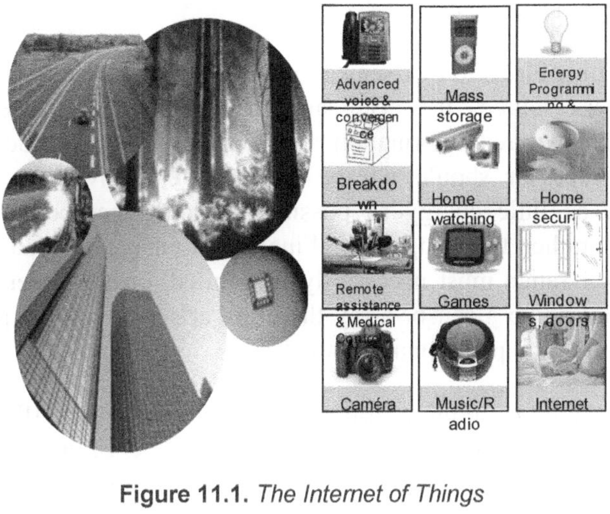

Figure 11.1. *The Internet of Things*

We begin by examining sensor networks, RFID interfaces and Near Field Communication (NFC) technology, which is increasingly used in the connection of "things" to the Internet. Then, we will give some ideas about networks experiencing extremely rapid growth: medical sensor networks, which potentially affect billion humans and therefore a huge potential market. Then, we will have a look to Fog networking which is a Cloud solution to connect things. Finally, we will introduce the different solutions for connecting things to the Internet.

11.1. Sensor networks

A sensor network is defined as a set of interconnected sensors, each sensor being connected through a transceiver. Sensor networks represent a new generation of networks with specific properties, which do not fall under conventional architectures. However, the arrival of the Internet of Things has changed the vision of sensor networks, which can of course continue to be a closed set, but also open up the Internet. Sensors can be very different from each other. For example, quite particular sensors are from so-called smart dust. This dust, which is almost invisible, has radio equipment in addition to the internal sensor-related treatment. Other sensors are coming from the home environment like security, temperature control, light control, telecommunications, etc.

The miniaturization of sensors poses problems of communication and power resources. It is necessary that the sensor is smart enough to gather the required information and send it wisely. In addition, the sensor processor should not be used too extensively in order to consume as little energy as possible. It must incorporate reactive elements rather than cognitive. Finally, to ensure a good rate, the transceiver range must necessarily be limited, on the order of about 10 m. Setting up a sensor array therefore arises of routing problems, error control and power management.

From the communication point of view, the IP protocol environment is too heavy and causes an excessive flow and consumption. Solutions that were derived from field bus or industrial real-time networks have a better compromise between efficiency and energy consumed. As hundreds of sensors may be placed per square meter, IPv6 addressing is a first obvious solution to the problem of managing addresses. However, the IP environment is heavy, and the implementation in these sensor networks of a protocol stack like TCP/IP or even UDP/IP is often impossible or is too much energy-consuming.

For now, the problems of security and quality of service are put into the background compared to consumer problems. An important field of research is in any case open to make effective and resilient sensor networks.

We can find networks using the ZigBee radio standard, which is pretty good as regards energy consumption. Wibree and 6LoWPAN are other solutions. Wibree is a very low-energy technology in a range of 10 m and a flow rate of 1 Mbps. This solution was developed by Nokia to compete with both ZigBee and Bluetooth. Yet, it was integrated in Bluetooth in 2009 to produce a product called Bluetooth Low Energy (LE).

The 6LoWPAN networks (IPv6 over Low power Wireless Personal Area Networks) come from an IETF working group. The objective is clearly to enable continuity of the IP protocol on less powerful machines and with limited electric power. The use of the IPv6 standard to obtain a very large number of addresses to large sensor networks is problematic. Indeed, the 16 bytes occupied by the sender address added to the 16 bytes of the receiver address require large control fields resulting in a poor utilization of the radio link to transport management information. This can become problematic because of the little power available within the sensor. ZigBee, however, limits the length of this frame to 127 bytes, which can also cause problems if the information to be transported from a sensor is long.

Sensor networks can form mesh networks, and in this case they need a routing protocol. Using a protocol such as IEEE 802.11s associated with IPv6 addresses would be catastrophic for the longevity of the battery inside the sensors. For this reason, the current proposals are much simpler, with protocols like LOAD (6LoWPAN Ad-hoc Routing Protocol), which is a simplification of AODV routing scheme, DYMO-Low (Dynamic MANET On-demand for 6LoWPAN), a simplification of DYMO coming from the MANET working group, and Hi-Low (Hierarchical Routing over 6LoWPAN), which has a hierarchical addressing. These proposals come from various protocols of the IETF standardization and thus ad-hoc networks, but not taking into account all options.

Another important feature of sensor network protocols relates to the discovery service, which must allow starting of the network automatically. The IETF also plays an important role in this area by proposing several solutions such as LoWPAN Neighbor Discovery Extension. This protocol is a reduction of the standard Neighbor Discovery and may be applied to broadcast and multicast management.

A particular sensor network is proposed by smart dust, which aims to develop nanotechnology sensors and link them together by an ad-hoc or mesh-type network. Smart dust is held in less than a cubic millimeter space, hence its name dust. In this dust are all the components necessary to realize a communicating computer: a processor, memory, radio, battery, etc.

The main problem here is still saving energy when performing the functions of the sensor. In particular, the network portion must carry communications expending very little power. The University of Berkeley has designed an operating system and specific protocols, called TinyOS and Tiny protocol. The TinyOS has been written in a simplified C language called NESC, which is a kind of dialect designed to optimize memory usage.

11.2. RFID

RFID has been introduced to achieve the identification of objects, hence its other name electronic tag. Electronic tags are interrogated by a reader, which allows us to retrieve the identification information.

Electronic tags are used in many applications, ranging from animal tracking tags to the identification of different products in a store.

There are two main types of RFID tags: passive tags and active tags. Passive tags do not have any energy resource. They are illuminated by a reader which provides an electromagnetic field sufficient to generate an electric current for transmission by a radio wave of the bits stored in a memory constituting the EEPROM RFID. A passive tag is shown in Figure 11.2. The antenna must have an architecture that enables it to receive the electromagnetic field of the reader and make this identity.

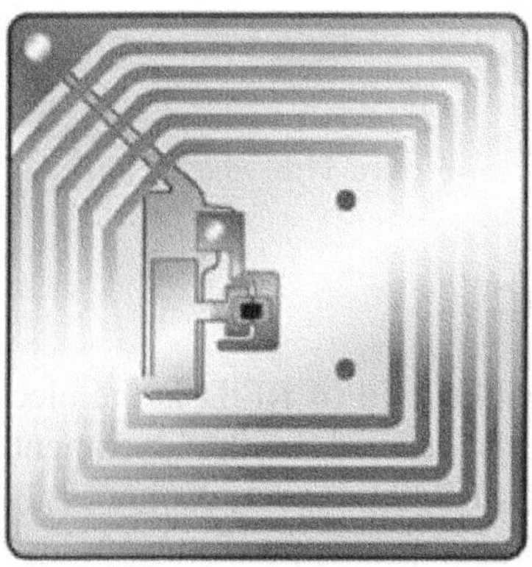

Figure 11.2. *Passive RFID*

A passive RFID can be very small. The necessary equipment is limited, the size of one-tenth of a millimeter is sufficient. The price of an electronic tag depends on the number of items manufactured in a single pass. It is possible to a cost of go down to a cost of five cents.

Active RFID tags have a power source in the component. The first major advantage of these tags is the quality of the transmission. A session can be opened between the reader and the RFID so that a retransmission can be carried out automatically in case of problems. Another advantage is the transmission with a range of several meters between the RFID and the reader, instead of a few centimeters. A disadvantage may be the lifetime of the battery. However, for standard use of some readings per day, it is possible to exceed 5 years. Active RFID can have more memory for storage of attributes associated with the value of identity. An active RFID is shown in Figure 11.3.

Figure 11.3. *Active RFID*

11.2.1. *Using RFID*

A well-known application of RFID is the electronic passport. The electronic passport is defined in the document International Civil Aviation Organization 9303 Part 1, Volumes 1 and 2 (6th edition, 2006). The electronic tag component contains information that is printed on the passport itself and a digitized photograph of the passport owner.

Transport titles form a second application of RFID. For example, a large number of metro or subway tickets contain an electronic tag that stores a set of information including the date and place of purchase. More sophisticated solutions are implemented in some public transportations, where the label becomes active and allows use of the same ticket several times.

Highway toll barriers also use this solution of active RFIDs, with ranges of a few meters. The active tag subtracts the toll from the amount of money stored in the memory. Similarly, toll gates to the lifts in many ski resorts use active RFID.

Another fairly immediate application is the monitoring of cars to detect stolen cars when passing near a reader. Finally, one of the most popular applications concerns inventories and purchases in shops. Inventories can be performed more often and with less error. Purchases placed in a basket can similarly be read by the reader. This provides a great simplification in the procedure for payment of purchases at a store.

The frequencies of transmissions of the RFID are shown in Table 11.1. They are determined by local or regional standardization bodies.

RFID frequencies	*Comments*
125 kHz (LF)	Lowest possible frequency for a relatively effective passive RDIF
13.56 MHz (HF)	The standard frequency used for passive RFIDs
400 MHz	Frequency used in specific cases, such as in the tracking of stolen cars
865–868 MHz (UHF)	Frequency band used in Europe
902–928 MHz (UHF)	Frequency band used in North America
2.4–2.4835 GHz	ISM radio band widely used for RFID applications

Table 11.1. *RFID transmission frequencies*

11.2.2. *EPC global*

RFID aims to give the object an identity to which they are backed. Standardization of this identification was carried out by the EPC global consortium. Two generations are available: EPC Gen1 and EPC Gen2. We will focus especially on this second generation released mid-2006, which has allowed the RFID to become an industrial technique.

EPC Gen2 is the acronym for UHF EPC global Class1 Generation 2. This specification was released in its version 1.1 in May 2006. The solution supports the protocol between the reader, on the one hand, and RFID and identification, on the other hand. The objective of the protocol is to read, write and kill an RFID under the condition that readers sold by any supplier are interchangeable.

The reading procedure is defined using a system based on time slots (slots) provided with a collision avoidance system. Indeed, a player can simultaneously turn on a large number of RFIDs. Simultaneous readings of different labels would result in collisions. A signaling is used to set the frequency, the encoding used (DSB-ASK, SSB-ASK and PR-ASK) and the flow channel. The collision avoidance system allows each simultaneous reading to keep only half of the objects that can transmit again in the next reading process. After a certain number of collisions, one RFID transmits successfully. The algorithm is designed such that each RFID is read in turn. The playback speed can reach 640 Kbps.

The identity of the object is determined by the EPC global Electronic Product Code. The Gen1 uses 96 bits, while the Gen2 switches to 256 bits. An example of this identification to the Gen1 is shown in Figures 11.4 and 11.5 which show the field identification in Gen2. This solution is much more complex because it uses intermediate filters that determine the lengths of the following fields. Note that the serial number password passes from 36 to 140 bits, allowing for a given thing to never return to 0 before thousands of years.

Type-1 Electronic product code			
8 bits	EPC Manager: 28 bits	Object class: 24 bits	Serial number: 36 bits

Header

Figure 11.4. *Structure of the electronic product code Gen1*

Type-2 Electronic product code					
8 bits	3 bits	3 bits	Company prefix: from 20 to 40 bits	Object reference: from 24 to 40 bits	Serial number: 140 bits

Header Filter value Partition

Figure 11.5. *Structure of the electronic product code Gen2*

11.2.3. *RFID security*

Security is a serious problem in the world of RFIDs. Indeed, reading a passive RFID can easily be done with almost any reader. Similarly, privacy may be undermined by the tracking and tracing of all that relates to an individual. Solutions are available, such as encryption of the identifier in the label or a change in the value of the label after each reading. These solutions are based on a middleware capable of interpreting the labels' values or tracking changes in value.

Active tags can open an authentication session for an exchange with the reader, which then acts as an authentication server. In this case, an electronic circuit in the tag must be able to encrypt a text issued by the authentication server. However, the key length is so small that the possibility of breaking after a series of authentication attempts is not negligible. Many proposals have been carried out using the anti-collision algorithm which allows us to serialize reading labels, for authentication.

11.2.4. *Mifare*

Mifare is the most widespread of contactless smart cards, with four billion cards in circulation in the world. The name comes from the company that created the card: Mikron. The name of the card was Mikron FARE. Mikron was acquired by Philips, which transferred ownership of this technology to its subsidiary NXP. There are two subtypes of Mifare cards: Mifare Classic, which uses only part of the commands of Mifare, and the NXP card, which implements the entire set of commands.

The cards communicate with a reader at a distance of less than 3 cm. This gives some security to the communications, which remain localized. However, attacks were carried out by very specific readers placed a few meters away. To be sure of the communication, it is better to encrypt data before transmission.

These cards are not very powerful in order to keep the cost to an extremely low level. Mifare Classic has a serial number of 4 or 7 bytes and 512 bytes of memory storage up to 4 Kbytes. The cheapest version is the Mifare Ultralight, which generally serves as a ticket.

Mifare T are cards using secure elements of the same type as a contact smart card and generally perform the same type of service.

The most advanced is the DESfire Mifare card, which has a larger storage memory and an advanced operating system, as well as an AES and 3DES encryption.

Mifare cards are used for many applications. The most common applications use a serial number, which triggers an action if that number is accepted by the reader. This action can be, for example, the reading of an identifier stored in the card or data storage. For the first application, we can use a number of 4 or 7 bytes. In the second application, the identifier must be inserted. The identifier may be encrypted as long as the tag has a secure element capable of performing cryptographic calculations.

11.3. Near-field communication

The NFC standard is a particular case of RFID communication. It enables us to communicate with a RFID reader at a distance of about 3 cm. This is a contactless communication with, for example, a smart card or a secure microcontroller.

Applications of this type are many and varied. The most classic is payment on mobile phones. With this mobile, a customer can recharge his or her account by a phone call or a low-speed data transfer to a server. Once the account is recharged, the mobile is a key to pay for a range of services. To use the Metro, for example, the client has just to move the mobile near the receiver to validate the radio communication for buying the ticket.

The flow rates of NFC communications are quite low: 106, 212, 424 and 848 Kbps. The frequency range corresponds to 13.56 MHz. The standards are derived from the ISO/IEC. The NFC Forum was created by Philips and Sony.

The security of the communication is ensured by the very small distance between the transmitter and the receiver and by the presence of the smart card. However, despite the distance of 3 cm, attacks were carried out in 2012 to listen to the NFC interface remotely. Therefore, it is advisable to encrypt the communication.

11.3.1. *Mobile key*

Car, home or hotel doors need keys that can be included in a smartphone as mobile keys. One advantage of these keys is that they are mobile and can be transmitted. A friend who is waiting outside your door can recover the key through an Internet connection. In the same way, mobile keys can simplify the distribution of keys for a car rental company. We will describe such an application through the opening of a hotel room by a smartphone.

Mobile keys can be stored on a secure server in a datacenter or in the Cloud. The smartphone, using a secure microcontroller, can store the key. Finally, the door of the room is an NFC lock. So, we have the environment depicted in Figure 11.6.

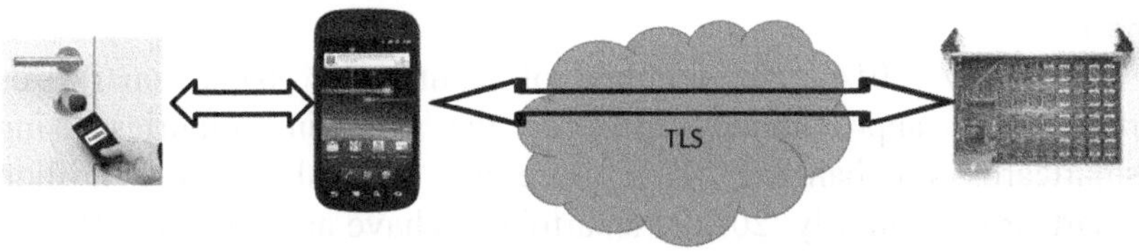

Figure 11.6. *The key to the mobile environment*

First, the room is booked through the smartphone. As indicated above, the smartphone must contain a secure element, that is to say, a location that cannot be accessed from the outside. The secure element can be of different types depending on the nature of the device (smartphone, laptop or tablet PC). This may be a SIM card from an operator, but the security provider must have the permission to add software to the smartcard. The secure element can also be an embedded SIM card, which is the equivalent of an operator SIM card but embedded in the smartphone by the manufacturer of the smartphone. This is mostly a NXP component. The secure element can also be embedded in a memory card (for example, an SD card). Finally, as shown in Figure 11.6, the secure element can be an intermediate smartcard (for example, a loyalty card of a hotel chain or an airline company) that communicates with the mobile through the NFC interface. The smartphone is seen as a modem allowing the external smartcard to communicate with the server where the keys are stored. This server is generally in the Cloud. The NFC door may be opened by the mobile key retrieved on the smartphone or the external NFC card.

Upon arrival in the vicinity of the hotel, detected by the GPS of the smartphone, the hotel transfers the key of the room into the secure element of the smartphone or directly into the external NFC smartcard. Then, the smartphone has to be moved close to the lock of the door to open it.

11.3.2. *NFC payment*

In early 2014, the number of contactless bank smartcards represented approximately 20% of the total in countries using smartcards for banking. This represents several hundred million smartcards. Similarly, 20% of smartphones have an NFC chip. Before the arrival of NFC, other payment systems had been tried, such as payment by SMS, direct payment model and different solutions using WAP technology.

Under the NFC, the mobile must be moved a few centimeters from the payment terminal. Two main areas have been set up for this prepayment and direct payment. In the first case, the terminal is seen as an electronic wallet: for each transaction, money is taken from the wallet. Payment for a very low maximum amount can be compared to the first axis. The second case requires more security since payment is made directly for the account of the user, for sums that are far greater generally. There are four possible models for mobile payments:

– Operator-centric model: the mobile operator deploys only its mobile payment services. It must provide for this independent mobile wallet account for the user. This case is unusual because the operators are not related to the major international payment networks. Mobile network operators must manage the interface with the banking network to provide advanced services to mobile payment. Pilots using this model have been launched in developing countries, but they cover only a small part of the many use cases of mobile payment services. Payments are generally limited to a transfer or modest sums.

– Bank-centric model: the bank deploys mobile payment applications associated with client devices and requires merchants to possess the equipment required to fulfill the sales. Mobile network operators are used as the transport network, with the quality of service and security required for payment functions.

– Collaboration model: this model involves collaboration between banks, mobile operators and a trusted third party.

– Peer-to-peer model: the mobile payment service provider acts independently of financial institutions and mobile network operators.

Google, PayPal, GlobalPay, etc., use a Cloud-based approach to make a mobile payment. This approach places the mobile payment provider in the middle of the operation, which involves two separate steps. First, a payment method associated with the cloud is selected and the payment is authorized via NFC. At this stage, the payment provider automatically covers the cost of purchases. In the second stage, another transaction is necessary: the payment provider must refund the payment and fees from the bank of the user.

11.3.3. *The Internet of Things in a medical environment*

The Medical Internet is booming, with 7 billion people to accompany, diagnose and treat. The ultimate idea is to continuously monitor individuals to advise or alert them to problems detected through a sensor. Every human being will in the future be connected continuously to a Cloud, which will analyze the data from the sensors disposed within and on the body surface. The power of the Cloud will allow us to analyze, compare and diagnose the data set stored. Essential features in this universe are security, privacy, that is to say, data privacy, and very low power consumption of the sensors, which must operate for months without the need to change the batteries. This environment is being set up gradually. We are going to describe the main advances from the network point of view.

The first element to develop relates to the adaptation of the sensors to the body through the body area network (BAN). An example of a BAN is depicted in Figure 11.7.

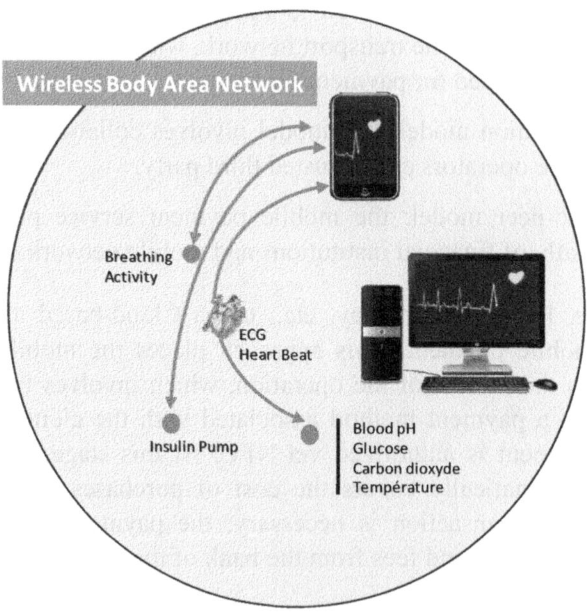

Figure 11.7. *A BAN (body area network)*

It is the IEEE 802.15.6 working group that deals with normalization. The main standards cover several frequencies and several technologies:

– 402–405<MHz for In-In-Out communications, that is to say, internal to the body and from the inside to the body surface;

– Ultra Wide Band (UWB), which operates on the 2.4 GHz and 3.1–10.6 GHz bands. UWB was described in Chapter 7;

– between 15 and 50 MHz for communications internal to the body;

– Human Body Communications, which operates at 13.5, 400, 600 and 900 MHz for communication with the body surface. This communication uses the conductivity of the human body. The intensity of the electric current is extremely small, of the order of a nanoampere.

To interconnect the BAN networks with the Cloud, several solutions are considered: either by LTE, or 4G, or a new generation of WiFi network: the FCC wants to develop a specific band for medical WiFi. But this new WiFi must ensure a security much stricter than the current WiFi security. The bandwidth is 40 MHz with a very low power consumption technology. The band for such external connections would be between 2,360 and 2,400 MHz, i.e. just below the current WiFi band.

Other research, which is progressing well, concerns the use of embedded sensors in "smart dust" capable of circulating in the blood. Analyses are regularly transmitted to external sensors, which themselves transmit the data to the Cloud.

11.4. The Internet of Things in the home

The home is a place where there are a large number of very different sensors from different backgrounds: information technology, telecommunications, medicine and electronics. We see, for example, Internet beds that have sensors capable of detecting movements and words during the night in order to establish a sleep diagnosis. Doors, household appliances items, security systems, etc., are themselves increasingly filled with diverse and varied sensors. Another example comes from the presence sensors that detect whether someone is in a room or not to shut down or automatically turn on the lights. A forecast predicts a hundred sensors per home by around 2020.

A major battle is engaged to determine the connection technology to be used by all these sensors in the home. The basic network is Ethernet, wired or WiFi. As a result, most of the equipment will be connected by an Ethernet jack or through a WiFi card. Connections can also use Bluetooth, 6LoWPAN, ZigBee and NFC. A communications controller should allow us to customize all connections in ensuring adequate security, availability, reasonable energy consumption and the lowest possible cost.

The problem comes from the next generation of Home Gateway, which pushes the management of IP in the end machine. For TVs, phones and other devices rather important, the presence of an IP component does not pose a problem: that is what we find in IPTV and IP telephony. However, for a sensor or a very inexpensive device, it is either impossible or too costly to introduce an IP component and to have an intermediate machine such as a controller. The controller can also be integrated into the Home Gateway.

In conclusion, the home contains a large number of "things" that can be connected. The Home Gateway is an essential piece of equipment because it can manage, possibly via a controller, not only communication, but also security and QoS for the home.

11.5. Fog networking

The Fog extends the Cloud to be closer to the things that produce and act on IoT data (Figure 11.8). These devices, called fog nodes, can be deployed anywhere with a network connection: on a factory floor, on top of a power line, alongside a railway track, in a vehicle or on an oil rig. Any device with computing, storage and network connectivity can be a fog node. Examples include industrial controllers, switches, routers, embedded servers and video surveillance cameras.

The most time sensitive data are analyzed on the fog node closest to the things generating the data. In a Cisco Smart Grid distribution network, for example, the most time sensitive requirement is to verify that protection and control loops are operating properly. Therefore, the fog nodes closest to the grid sensors can look for signs of problems and then prevent them by sending control commands to actuators.

Data that can wait seconds or minutes for action are passed along to an aggregation node for analysis and action. In the Smart Grid example, each substation might have its own aggregation node that reports the operational status of each downstream feeder and lateral.

Data that are less time sensitive are sent to the Cloud for historical analysis, big data analytics and long-term storage (see sidebar). For example, each of thousands or hundreds of thousands of fog nodes might send periodic summaries of grid data to the Cloud for historical analysis and storage.

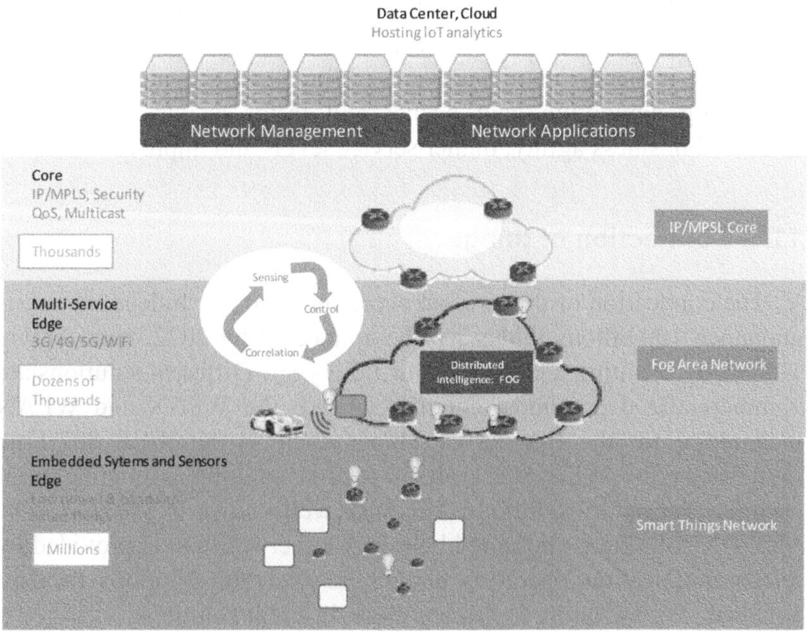

Figure 11.8. *Architecture of the Fog networking*

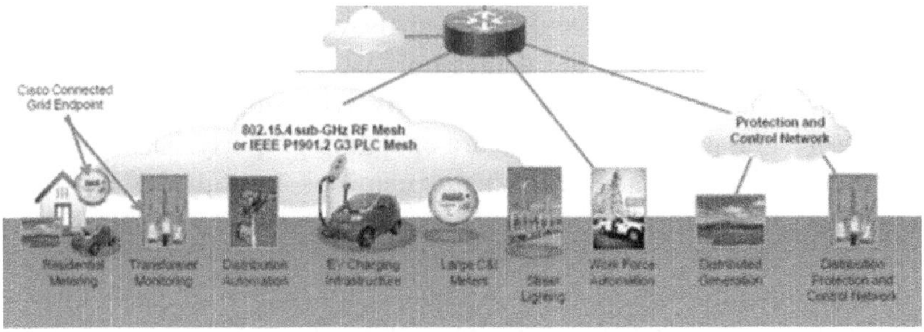

Figure 11.9. *Fog networking*

Fog computing gives the Cloud a companion to handle the two exabytes of data generated daily from the Internet of Things. Processing data closer to where it is produced and needed solves the challenges of exploding data volume, variety and velocity.

Fog computing accelerates awareness and response to events by eliminating a round trip to the Cloud for analysis. It avoids the need for costly bandwidth additions by offloading gigabytes of network traffic from the core network. It also protects sensitive IoT data by analyzing it inside company walls. Ultimately, organizations that adopt Fog computing gain deeper and faster insights, leading to increased business agility, higher service levels and improved safety.

11.6. Connection of things

The connection of things is an excellent business. Indeed, forecasts announce 100 billons things will be connected by 2022, representing 100 B$ by counting a connection at 1 $ a year. Different solutions are commercialized or almost commercialized: the WPAN and WLAN technology for the Internet of Things such as the IEEE 802.11ad (WiGig) or the IEEE 802.11ah. A second solution is coming from specific proprietary solutions. The most well-known is SIGFOX. The SIGFOX solution is presented below. Another solution is provided by telecommunication operators adding narrowband solutions to their antennas. We are going to present these different solutions.

11.6.1. *Specific proprietary solution: SIGFOX example*

The SIGFOX network technology allows for a very long range between objects and transceivers, which enables the network to easily cover large areas and reach underground objects. The SIGFOX technology allows for better penetration than traditional cellular networks such as GSM, while still ensuring high reliability and radically lower energy consumption. The SIGFOX network operates in the unlicensed ISM radio bands. The exact frequencies can vary depending on national regulations. The different bands are described in Figure 8.14.

In order for an object to send messages using SIGFOX, it needs to integrate a compatible and certified modem. The SIGFOX modems are based on standard hardware chips on which the SIGFOX software stack is installed. SIGFOX uses a patented radio technology based on what is referred to as Ultra Narrow-Band (UNB). The SIGFOX UNB radio communication uses the ISM bands, which are free to use without the need to acquire licenses, to transmit data over a very narrow spectrum to and from connected objects.

The SIGFOX network is not targeting specific industries and can be used in virtually any context in which there is a need for a low throughput connectivity solution. The definition of low throughput on the SIGFOX network can be characterized as follows:

– up to 140 messages per object per day;

– payload size for each message is 12 bytes;

– wireless throughput up to 100 bits per second.

In its current state, each SIGFOX base station can handle up to a million connected objects, but the network is easily scalable to handle more objects, by simply increasing the density of base stations. The density of the cells is based on an average range of about 30–50 km in rural areas and 3–10 km for urban areas. Distances can be much higher for outdoor objects where messages can be in line of sight.

11.6.2. *LoRa*

LoRa is the usual name to describe a solution provided by telecommunication operators to enter the Internet of Things. LoRa gains its name from the fact that this solution is able to provide Long Range communications using very low power levels. Indeed, LoRaWAN defines the communication protocol and system architecture for the network, while the LoRa physical layer enables the long-range communication link.

The protocol and network architecture have the most influence in determining the battery lifetime of a node, the network capacity, the Quality of Service, the security and the variety of applications served by the network. We describe in Figure 11.10 the LoRa architecture and in Figure 11.11 the LoRa network.

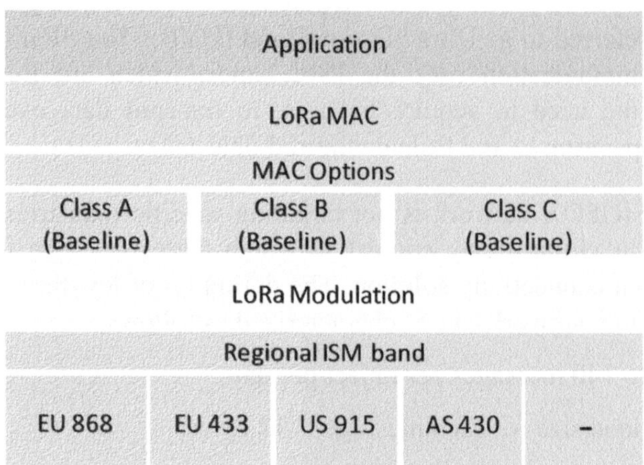

Figure 11.10. *The LoRa architecture*

End-devices serve different applications and have different requirements. In order to optimize a variety of end application profiles, LoRaWAN utilizes different device classes. The device classes trade off network downlink communication latency versus battery lifetime. In a control or actuator-type application, the downlink communication latency is an important factor. The three classes are depicted in Figure 11.12.

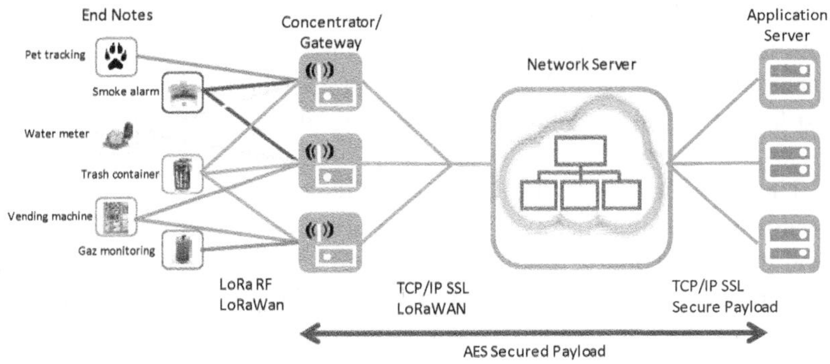

Figure 11.11. *LoRa network*

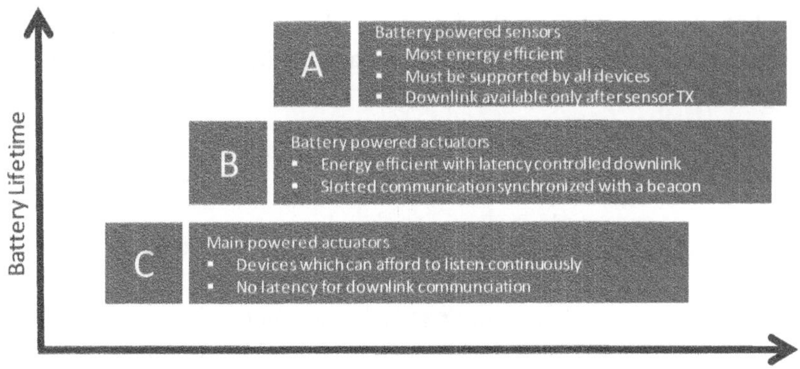

Figure 11.12. *The three classes of LoRa*

LoRa is the physical layer of the wireless modulation utilized to create a long range communication link. Many legacy wireless systems use frequency shifting keying (FSK) modulation as the physical layer because it is a very efficient modulation for achieving low power. LoRa is based on chirp spread spectrum modulation, which maintains the same low power characteristics as FSK modulation but significantly increases the communication range.

Chirp spread spectrum has been used in military and space communication for decades due to the long communication distances that can be achieved and its robustness to interference, but LoRa is the first low cost implementation for commercial usage.

The advantage of LoRa is in the technology's long range capability. A single gateway or base station can cover entire cities or hundreds of square kilometers. Range highly depends on the environment or obstructions in a given area, but LoRa and LoRaWAN have a link budget greater than any other standardized communication technology. The link budget, typically given in decibels (dB), is the primary factor in determining the range in a given environment. With a minimal amount of infrastructure, entire countries can easily be covered.

Let us present a little bit more detail on the European and American solution using to different parts of the spectrum as shown in Figure 11.13.

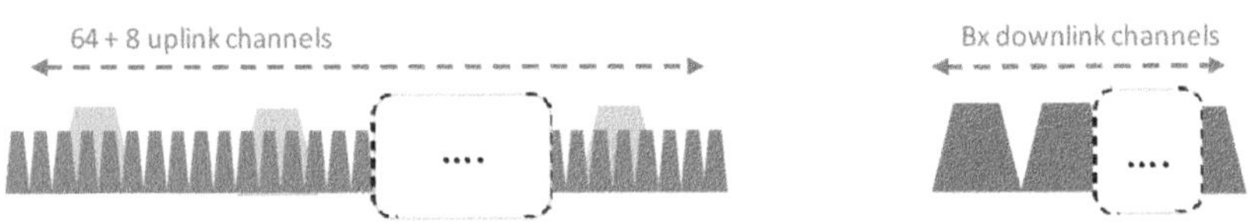

Figure 11.13. *The ISM band*

LoRaWAN for Europe defines 10 channels, eight of which are multi data rate from 250 bps to 5.5 kbps, a single high data rate LoRa channel at 11 kbps, and a single FSK channel at 50 kbps. For North America, the ISM band is from 902 to 928 MHz. LoRaWAN defines 64, 125 KHz uplink channels from 902.3 to 914.9 MHz in 200 KHz increments. There are an additional eight 500 KHz uplink channels in 1.6 MHz increments from 903 to 914.9 MHz. The eight downlink channels are 500 KHz wide starting from 923.3 to 927.5 MHz.

It is extremely important to incorporate security in a LoRa network. LoRaWAN utilizes two layers of security: one for the network and one for the application. The network security ensures authenticity of the node in the network, while the application layer of security ensures the network operator does not have access to the end user's application data. AES encryption is used with the key exchange utilizing an IEEE EUI64 identifier. This is shown in Figure 11.14.

As a conclusion to this section, we show the global architecture of the LoRa solution in Figure 11.14.

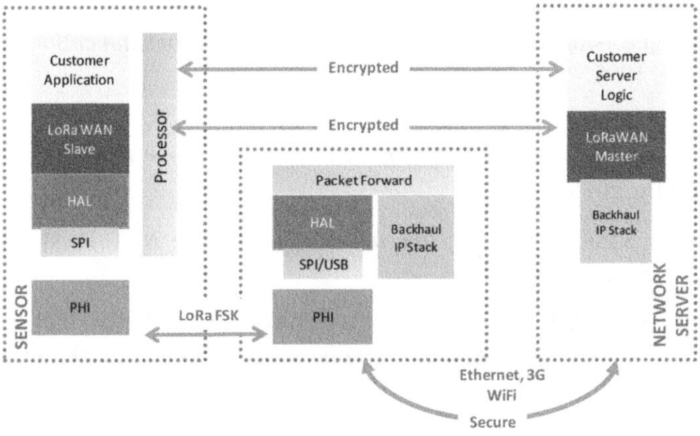

Figure 11.14. *The Lora solution*

11.7. Conclusion

The Internet of Things is becoming a reality due to RFIDs and sensors. The integration of these tiny devices is far from complete, as they are not powerful enough to support the TCP/IP protocol. To get there, very specific gateways have to be developed, which may include, on the one side, the connected things and, on the other side, the Internet.

11.8. Bibliography

[BAH 14] BAHGA A., MADISETTI V., Internet of Things (A Hands-on-Approach), available at www.Internet-of-Things-book.com, 2014.

[BRO 95] BRODSKY I., *Wireless: The Revolution in Personal Telecommunications*, Artech House, 1995.

[BUR 02] BURKHART J., *Pervasive Computing: Technology and Architecture of Mobile Internet Applications*, Addison Wesley, 2002.

[MAS 14] MADISETTI V., BARHGA A., *Internet of Things*, VPT Publisher, 2014.

[MUL 95] MULLER N.J., TYKE L.L., *Wireless Data Networking*, Artech House, 1995.

[PRA 98] PRASAD R., *Universal Wireless Personal Communications*, Artech House, 1998.

[ROW 15] ROWLAND C., GOODMAN E., CHARLIER M. *et al.*, *Designing Connected Products: UX for the Consumer Internet of Things*, 1st ed., O'Reilly, 2015.

[SAM 15] SAMUEL G., *The Internet of Things*, The MIT Press Essential Knowledge Series, 2015.

[SHO 06] SHOREY R., ANANDA A., CHAN M.C. *et al.*, *Mobile, Wireless, and Sensor Networks: Technology, Applications, and Future Directions*, 1st ed., Wiley-IEEE Press, 2006.

[STA 15] STACKOWIAK R., LICHT A., MANTHA V. *et al.*, *Big Data and The Internet of Things: Enterprise Information Architecture for A New Age*, Apress, 2015.

[WAH 15] WAHER P., *Learning Internet of Things*, PACKT, 2015.

12

Ad Hoc and Mesh Networks

Radio networks with relay refer to communications that pass through intermediate equipment in the radio part. The packets are memorized and processed before retransmission to another device, always over the air. These networks are also referred to as multi-hop networks. When the relays form a mesh network in which each node is connected to several other nodes, they are called mesh networks. The access points forming the mesh network can be fixed or mobile. These nodes form the base network, in general belonging to an operator or a network provider. Customers of a mesh network are connected on the mesh routers, usually through another access card available in the mesh router. In a mesh network, the customers have no routing software, and they are connected as a hotspot.

An ad hoc network looks like a mesh network, except that the routing software is on the customer's terminal. In other words, there is no specific machine in an ad hoc network: the clients form the network. The big drawback comes from a very strong dependency between routers and client devices. However, these are networks without infrastructure since they include only the client terminals. They are sometimes called "Wild networks" because they can be performed independently without any operator. However, it is illusory to think that a customer could communicate with a client in a town far away through intermediate nodes representing several hundreds or even thousands of terminals. The reason for this problem comes from the management of routing tables, which become huge and highly dynamic since the client terminals move constantly.

The standardization of ad hoc networks was carried out by the IETF working group MANET (Mobile Ad hoc NETwork).

Another radio network with relay is formed by the Vehicular Ad hoc NETwork (VANET). These networks are rapidly expanding in the automotive industry. This solution allows us to connect customers to one another within a vehicle, the vehicle may be a bus, a train or an airplane.

We have also consolidated in this chapter satellite networks using satellites as relays. These satellites increasingly often have even routing processing capabilities in satellite constellations, which allows us to include them in the radio networks with relays.

12.1. Ad hoc networks

Mobile Ad hoc NETworks (MANETs) form an important class of radio networks with relays. The infrastructure is composed only of the stations themselves playing the role of transmitter, receiver and router. The routing enables the passage of information from one terminal to another, without these terminals being connected directly. An ad hoc network is shown in Figure 12.1.

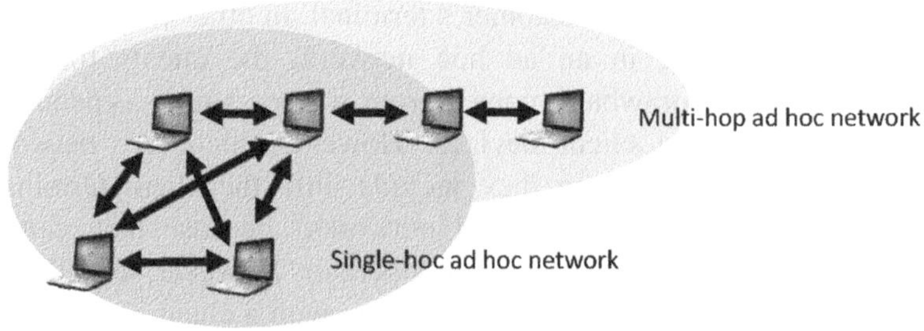

Figure 12.1. *An ad hoc network*

Contrary to appearances, ad hoc networks are quite old. They aim to achieve a communication environment that unfolds without any infrastructure other than the mobile equipment itself. In other words, the mobile equipment can act as a gateway to allow communication from one machine to another. Two pieces of mobile equipment moving too far away from each other to directly communicate can find intermediate mobile equipment able to act as a relay. The main problem caused by this network comes from the definition of the network topology: how to identify the neighboring nodes and how to move from one node to another node? Two extreme solutions can be compared. The first is that of an ad hoc network in which all nodes can communicate with all the others involving a long-range transmitter. In the second solution, instead, the radio range is as short as possible: in a communication between two nodes, it usually goes through several intermediate machines. The advantage of the first solution is the security of the transmission, since the communications can go directly from the sender to the receiver, without depending on an intermediate device. Network throughput is minimal: the frequency cannot be reused. In the second case, if a device fails or is turned off, the network may be cut into two separate subnetworks without communication to one another. Obviously, in this case, the overall throughput is optimized, since there may be a high-frequency reuse.

The access techniques are of the same type as in mobile networks. However, due to the fact that all laptops are mobile, new properties should be made for the management of user addresses and routing control.

The usual solution developed for ad hoc networks takes the IP environment to define the routing scheme. Mobile equipment which acts as a gateway – usually all mobile equipment – implements a router, so that the problems arising essentially depend on the Internet routing schemes.

The benefits of ad hoc networks are the simple extensions, the physical coverage and the cost. However, a certain number of obstacles have to be overcome due to the mobility of nodes, particularly regarding the Quality of Service and security.

MANET is an IETF working group that is concerned with the standardization of ad hoc protocols running IP. This group is linked on conventional Internet protocols and has improved them so that they can work with mobile routers. Two major protocol families were defined: reactive protocols and proactive protocols:

– Reactive protocols. The terminals do not maintain a routing table but when a transmission happens a routing has to be performed. In this case, the scheme primarily uses flooding techniques to identify the mobile equipment that can participate in the transmission.

– Proactive protocols. Mobile equipment tries to maintain a consistent routing table, even in the absence of communication.

Ad hoc networks are useful in many cases. They allow us to set up networks within a limited period of time, in the case, for example, of an earthquake or during a meeting with a large number of participants place. Another possibility is to extend access to a cell of a wireless network such as WiFi. A terminal located outside of a cell can connect to another machine located in the cell coverage area. The latter serves as an intermediate router to access the antenna of the cell.

Ad hoc networks raise many problems because of the mobility of all equipment. The main problem is the routing that requires us to forward packets from one point to another point in the network. One of the objectives of the MANET group is to propose a solution to this problem. Four major proposals have emerged, two of a reactive type and two of a proactive type. Among other issues, we find the safety, Quality of Service and mobility management during the call.

12.2. Routing

The main idea of an ad hoc network is that each wireless node can communicate directly with any other wireless node, without the use of any existing network infrastructure. Thus, the architecture is totally decentralized. Such networks are limited by the fact that each wireless node has to be within the transmission range of all the others, which involves small networks. The main problems in ad hoc networks are how to transmit in the air medium and how to share it. There exist two major link layer technologies responding to this: the IEEE 802.11 standard and the Bluetooth specifications.

Therefore, the architecture of an ad hoc network applies to small networks, due to the limited transmission range of wireless network interfaces. As shown in Figure 12.1, extending connectivity to a larger area requires a multi-hop ad hoc network architecture, in which each node assumes the role of a router and relays packets toward final destinations. Thus, network nodes cooperate to forward packets for each other to allow communication over multiple "hops" between nodes not directly within wireless transmission range of one another. But this characteristic is not sufficient to ensure network connectivity. In addition, packets must be directed through the network. In other words, a path has to be determined from the source to its destination. This is so-called routing. A simple way to perform it would be to enter static routes by integrating a fixed routing table in each node according to the topology. Obviously, as nodes in a multi-hop ad hoc network move about or join/leave the network and as wireless transmission conditions such as sources of interference change, this technique is not conceivable. Routing must result from a specific algorithm which calculates routes automatically and dynamically, that is to say from a multi-hop wireless ad hoc network routing protocol.

Before presenting the standardized routing protocols in section 12.4, we recall that multi-hop ad hoc networks have the following characteristics:

– dynamic topologies: nodes, which are able to forward packets, can move arbitrarily. This mobility makes the topology changing randomly and rapidly at unpredictable times;

– limited physical security: attacks, such as denial-of-service or eavesdropping, are often easier to lead in wireless networks than in wired ones. Ad hoc networks have at least the advantage of being totally decentralized;

– omnidirectional radio transmission: packets are broadcasted to all neighboring nodes. Then, simultaneous data exchange between nodes close to each other can be impossible;

– possible unidirectional links: because of interferences, fading or still limited transmission range of some wireless network interfaces, links can be unidirectional;

– bandwidth and energy constraints: these resources are quite limited. The former due to the intrinsic properties of radio interfaces and the effects of multiple access, noise, etc.; the latter due to the characteristics of wireless mobile nodes forming the multi-hop ad hoc network, which often rely on batteries. Routing not only consists of finding a route between two nodes, the goal is to calculate an optimized route according to one or several constraints (number of hops, maximum bandwidth, minimum end-to-end delay, etc.). Not designed for ad hoc particularities, classical routing protocols (e.g. RIP and OSPF) were inapplicable.

They had to be upgraded in order to be efficient and reliable, taking into account the wireless and decentralized architecture, reducing control traffic and managing mobility. Therefore, multi-hop ad hoc routing protocols present much originality, since the mobile nodes in the network dynamically establish routing among themselves to form their own network. Nevertheless, these protocols apply properties of classical routing schemes. Indeed, we can distinguish two techniques to calculate the best route:

– distance-vector routing: each node informs (periodically and at topology changes) its neighbors of the estimated direction and distance to any other node. The cost of reaching a destination is computed using various route metrics. The Bellman–Ford algorithm is then used to calculate routes. Distance-vector routing protocols have low computational complexity and message overhead. On the other hand, the algorithm has a struggle to converge on the new topology when network conditions change;

– link-state routing: each node broadcasts in the network the links state with its neighbors. Being acquainted with which nodes are connected to which other ones, every node ends by constructing a map of the connectivity to the network. Then, the Dijkstra algorithm enables us to calculate optimal routes. Instead of sharing routing tables with the neighborhood, nodes applying a link-state routing protocol flood the network only with their related connectivity. These two techniques, well known in wired networks with RIP and OSPF, are used in multi-hop ad hoc routing protocols (e.g. AODV and OLSR). Many other differences appear between routing protocols. For instance, a packet may contain in its header the complete path to follow in order to reach the destination (source routing) or only the next node to its destination (hop-by-hop). We will see that a major distinction in multi-hop ad hoc routing protocols concerns the time when a route is calculated. This defines two kinds of protocols: table-driven and on demand protocols.

12.2.1. *Ad hoc in the link layer*

Several technologies enable ad hoc networks, both in Wireless Personal Area Networks (WPAN) and Wireless Local Area Networks (WMAN). By definition, WPANs are small networks, limited to a few meters, with data rates up to several hundreds of Kbit/s. As for WLANs, their communicating range is larger (e.g. a building) and they can offer in the region tens of Mbit/s. A WLAN is obviously and above all a LAN. Thus, in spite of wireless characteristics (e.g. bandwidth and energy constraints), the requirements are the same: high data rates, full connectivity, broadcast capability, etc. We present in the following the two main link layer technologies for ad hoc networks: the IEEE 802.11 standard for WLANs and the Bluetooth specifications for WPANs. As detailed in the previous section, a "single-hop" ad hoc network implies that all the stations are within the same transmission area to be able to communicate.

The range of the network can be extended by adding network layer protocols, including routing mechanisms. Solutions leading to multi-hop ad hoc networks are described in the next section.

12.2.2. *Ad hoc mode in WiFi*

Proposed in 1997, the IEEE 802.11 standard (commercially called WiFi, the acronym for Wireless Fidelity) defines the two first layers of the OSI model for a wireless local area network: the physical layer and the link layer. While three different physical layers are proposed, the link layer is unified. Indeed, to transmit data between nodes, there can be used either: infrared (IF) pulse position modulation, direct sequence spread spectrum (DSSS) or frequency-hopping spread spectrum (FHSS). Frequencies are in the Industrial, Scientific and Medical (ISM) band, ranging from 2.4000 to 2.4835 GHz. This standard allows data rates of equal to 1 or 2 Mbit/s. It was then extended by several task groups, which are designated by letters. In 1999, the extensions IEEE 802.11a and IEEE 802.11b were proposed in order to improve data rates: up to 54 Mbit/s in the 5 GHz band for the former and up to 5.5 or 11 Mbit/s in the 2.4 GHz band for the latter (with backward compatibility). In 2003, the IEEE802.11g extended the IEEE 802.11b to support 54 Mbit/s in the 2.4 GHz band, by applying the same OFDM-based transmission scheme as IEEE 802.11a. Nowadays, there exist around 30 amendments, offering higher throughput, security services, handover capabilities, etc. Two operational modes are specified in the standard:

– infrastructure-based mode: in this mode, all the communications between nodes have to go by a specific node: the Access Point (AP). The network area is then determined by the transmission range of the AP. The set of nodes around the AP is called the Basic Service Set (BSS). To extend the network, it is possible to merge several BSSs. For this, the AP can be connected to another AP, generally with a wired network such as Ethernet;

– ad hoc mode: no AP is required in this mode, the network being totally distributed. Nodes can communicate if they are within their respective transmission range. They constitute an Independent Basic Service Set (IBSS). Moreover, the Medium Access Control (MAC) layer of the IEEE 802.11 standard defines a contention-free service, called the Distributed Coordination Function (DCF) and based on a multiple access method. As an optional access method, the standard defines the Point Coordination Function (PCF), implemented on top of the DCF and based on a polling scheme. Since there is no central authority to act as a point coordinator in ad hoc networks, the PCF cannot be readily applied in such environments.

In the DCF mode, a Carrier Sense Multiple Access with Collision Avoidance (CSMA/CA) is used. Hence, to transmit a packet, a node has to first listen to the channel. If it is idle, then the node can send its packet. Otherwise, that is to say if the node senses energy above a specific threshold on the medium, the transmission is deferred for a random interval, reducing the probability of collisions on the channel. More precisely, when the current transmission ends, the node wanting access waits a Distributed Inter-Frame Spacing (DIFS) before choosing a random backoff value. Then, this value is decremented only when the channel is idle. As soon as the value reaches zero, the node sends its packet. Despite this mechanism, collisions may occur. Contrary to the CSMA/CD used in the IEEE 802.3 standard, collision detection is not feasible in the IEEE 802.11 standard, due to the unique antenna. This hinders the node from sending and receiving simultaneously. Collisions are then detected in oblique ways: to confirm the correct reception of information, acknowledgment (ACK) packets are sent from the receiver to the sender. In order to send the ACK just after the received packet, another interframe spacing is defined: the Short Inter-Frame Spacing (SIFS). This one being shorter than the DIFS, ACKs are sent as a priority, preceding any other packet transmission.

Another problem occurs in wireless networks: the hidden node problem. Indeed, let us consider Figure 12.2 where node C wants to transmit a packet to node B, which is already receiving a packet from node A. As C and A are not within the transmission range of each other, when node C applies the CSMA protocol and so listens to the channel, it cannot detect the transmission between A and B. This creates a collision, scrambling data.

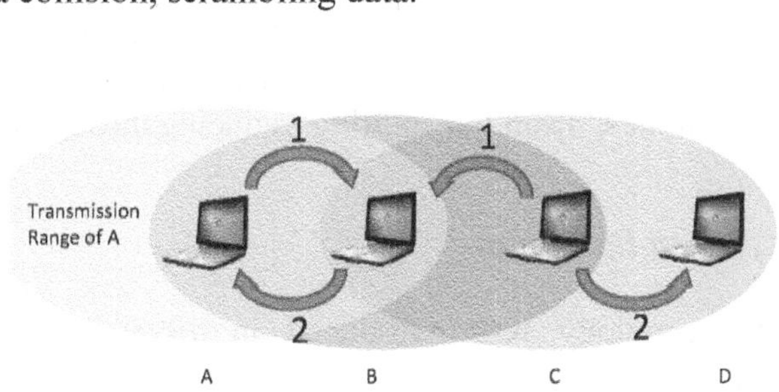

Figure 12.2. *Hidden node (1) and exposed node (2) problems*

To solve this problem, the physical carrier sensing mechanism of CSMA can optionally be supplemented by an additional mechanism, a virtual carrier-sensing one based on the exchange of Request To Send (RTS) and Clear To Send (CTS) packets between the sender and the intended receiver. The IEEE 802.11 RTS/CTS exchange is defined as follows: before sending packets, the source sends to the intended receiver a short packet, named RTS. If the RTS packet is well transmitted, the receiver replies (after a SIFS) by another short packet, named CTS. Then, the source knows that the destination is ready to receive the data frame. Information contained in both the RTS and CTS packets is the identity of the requesting node, this of the intended receiver and the total duration of the transmission, that is the time interval needed to transmit the RTS and CTS packets, the data frame and the related acknowledgment. Therefore, all the nodes within the transmission range of either the source or the destination know the overall time interval of the coming transmission. It is important to note that this mechanism partly overcomes the hidden node problem. Moreover, RTS/CTS may decrease throughput. This is the reason why an RTS threshold can be adjusted in order to apply this technique only for large data frame transmissions that are those considerably reducing network performances when collisions occur.

Another technique consisting of increasing the transmitting power of the nodes could solve the problem. Indeed, this enables the nodes to hear the hidden ones, which are no longer hidden. Unfortunately, this technique amplifies another well-known problem in wireless networks: the exposed node problem. As shown in Figure 12.2, when B is transmitting to node A, even if C wants to send packets to D (a transmission that would not cause a collision at A), the CSMA protocol hinders node C from transmitting, reducing throughput.

12.2.3. *Bluetooth link layer*

Bluetooth technology was introduced in 1998 by a specific working group, the Bluetooth Special Interest Group (SIG), including industry-leading companies such as Ericsson, Intel, Nokia, Toshiba, 3Com and Motorola. These specifications have been resumed and extended by the IEEE 802.15 working group, in charge of developing a family of communications standards for WPANs. The goal of this technology was a wireless communication system based on short-range radio at low-cost. The application field was clearly the PANs by making feasible wireless ad hoc transmissions between different equipment such as laptops, printers, keyboards, cellular phones, wireless handhelds, car radios, etc. Several Bluetooth specification versions have been released since the original one (version 1.0A) in 1999. The Bluetooth specification defines the lowest layers, especially the radio and the baseband ones. The radio layer defines how to transmit data over the air medium. Bluetooth technology operates in the 2.4 GHz ISM band, employs a fast frequency hopping spread spectrum (FHSS) radio technique and uses a binary Gaussian frequency-shift keying (GFSK) modulation. These techniques enable very low-cost implementations. The baseband defines how Bluetooth devices can communicate with each other. A Bluetooth network is formed in an ad hoc manner (without any infrastructure assistance). The basic single-hop ad hoc network is named a piconet in the Bluetooth parlance. The transmission range of a specific Bluetooth device depends on the transmission power. Three classes are defined: class 1 (100 mW), class 2 (2.5 mW) and class 3 (1 mW) radios. The network area spreads from 1 to 100 m, according to the transmit power class.

During the piconet establishment, one node is designated master and the other nodes slaves. A piconet contains at most seven active slaves (actively involved in communications). A node that is not the master or an active slave is either parked if it is registered with the master or in stand-by mode if it does not belong to a piconet. The role of the master is to arrange transmissions in the piconet, by polling the slaves. Hence, a slave can deliver a packet only if it has received authorization from the master. Moreover, all transmissions have to go by the master, even if the data exchange is between two slaves. A locally unique Active Member Address (AM ADDR) is assigned to each active slave by the master, in order to identify them.

To communicate, all the nodes belonging to a given piconet are synchronized to the same channel, which is defined as the sequence of the frequency hops followed (up to 1,600 per second). Indeed, transmissions are made on a slotted scheme. The duration of a basic slot is equal to 625 μs. During a slot, the carrier frequency remains unchanged. Frequency hopping occurs at the end of the slot. However, when a packet occupies multiple slots (three or five), no frequency change is applied. When hopping resumes, the next frequency is the same as the one that would have been chosen in single slot transmissions. According to the Bluetooth clock of the master, slots in a piconet are even or odd. The master and the slaves alternate transmissions in a Time Division Duplex (TDD) manner, the even slots are used by the master and the odd ones are used by the active slaves, knowing that a slave can transmit only if it has just received a poll packet from the master. A slave may belong to several piconets. In this way, multi-hop ad hoc networks can be created. Such a Bluetooth network, interconnecting several piconets, is called a scatternet. More precisely, a slave can be temporary separated from a piconet to become the slave of another one, and then come back to the initial situation. A slave can also become the master of another piconet. These situations are illustrated in Figure 12.3. Several solutions are proposed in the literature in order to form a scatternet.

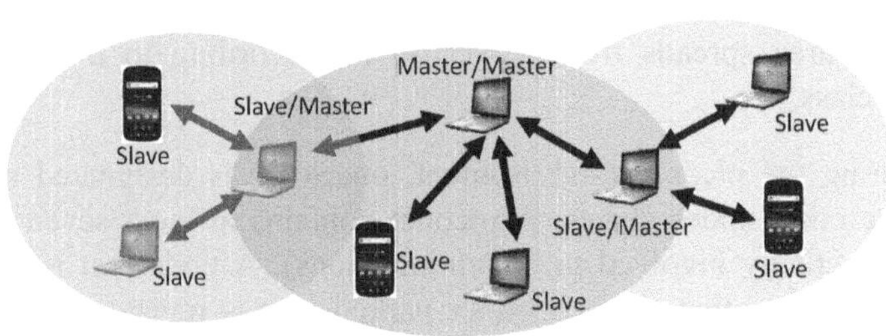

Figure 12.3. *Example of a Bluetooth scatternet*

12.3. Ad hoc routing protocols

Routing consists of finding a set of links which allow the transfer of the information from a source point to a destination. In ad hoc networks, links are mostly radio interfaces connecting two mobile nodes. The use of radio interface and mobile nodes makes the routing harder than in fixed networks because of many problems such as interferences, dynamicity, non-symmetric links, narrow band resources, non-deterministic access, etc.

For those reasons, applying the same routing protocols created for the IP networks is not possible and may create hard problems such as delivery divergence, high delay and small delivery rate of the information. A huge number of studies were done for creating or adapting protocols for ad hoc environments. Those protocols are divided into three sets: reactive, proactive and hybrid. The reactive protocols minimize the number of exchanged control information when nodes are idle and no traffic is active. They react only when a node decides to send information and at that time, the network starts a route discovery process that could be very expensive in terms of resources. On the contrary, proactive protocols exchange control information periodically in order to maintain information on the network topology and having information on routes for all the destinations. Finally, the hybrid protocols are a mixture of reactive and proactive.

At the IETF standardization process, the MANET working group was created in order to provide RFC documents on a reference protocol and the result was four experimental standards: two reactive (DSR and AODV) and two proactive (OLSR and TBRPF). Phase 2 was initiated and works are in progress to evaluate the reactive protocols toward one single protocol named DYMO and the proactive protocols toward one another single protocol which is OLSRV2.

The military was probably the first to require the use of ad hoc networks. Military operations are often very dynamic in nature and cannot be in most cases based on a fixed and pre-existing infrastructure. Communications between the various entities of a transaction will often be on a wireless while having no knowledge of the network formed by these entities.

The US Department of Defense was quickly involved in projects of advanced research and development in areas such as communication networks required by the military applications. In 1973, The Defense Advanced Research Agency, established by the U.S. Department of

Defense, initiated a research project on the technique of packet switching in wireless environments. This request is motivated by the need to provide network access to mobile entities and to establish communications in a mobile environment. The project would be based on the packet radio transmission developed at the University of Hawaii in the ALOHA project in the 1970s. Initially, the system could send data packets using radio to reach destinations located on one hop distance. The development of packet radio network (called later PRNet – Packet Radio Network) has enabled the radio multi-hop over an extensive geographic area. In Figure 12.4, the PRs represent nodes which are connected by radiolinks. All PRs can communicate directly by radio (Figure 12.4(a)) or by using intermediate PRs (Figure 12.4(b)). The latter is called multi-hop communications. If, for example, a communication must pass by three radio links, one speaks of a three hop.

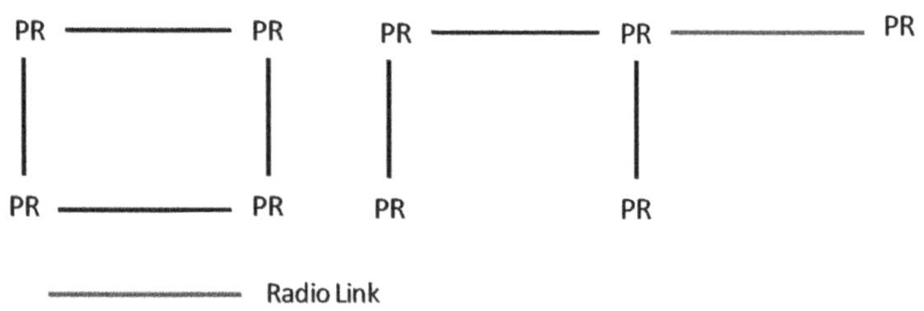

Figure 12.4. *Examples of PR-network*

The PRNet is the basis of ad hoc networks. The reader will find a complete description of packet radio networks in MANET which is a working group at the IETF standardization society. MANET was created at 1995 in order to provide RFC to propose routing protocols for ad hoc networks. At that time, only Hiperlan 1 could provide information forwarding in ad hoc environment. The latter was creating a layer 2 routing which is not compliant with other standards. The objective of MANET is to adapt the IP routing to work in ad hoc networks. From the long works and discussions, we standardize four protocols which belong to two different concepts: reactive and proactive.

12.3.1. *Reactive protocols*

Reactive protocols offer routes for the information flows only on demand. This means that the protocol will react to find a route when a source node needs to send information to a destination node. During the remaining time, the node is idle or active in participating in the forwarding process to serve other source nodes. The reactive protocols do not consume many resources when idle, however, to discover a route and since they do not own any topology information, they flood requests among the network and in this case consume a huge amount of the network resources. Examples of reactive protocols are AODV and DSR.

12.3.2. *Ad hoc on-demand distance vector*

AODV belongs to the reference of distance vector routing protocols. In these protocols, the mobile will know how to reach certain destinations, the smallest distance in number of hops and the next relay (hop). AODV is a reactive protocol; hence, mobile nodes do not retain any information for nodes not concerned by the active traffic information flows. After the traffic forward is achieved, the routing information is maintained for a timeout.

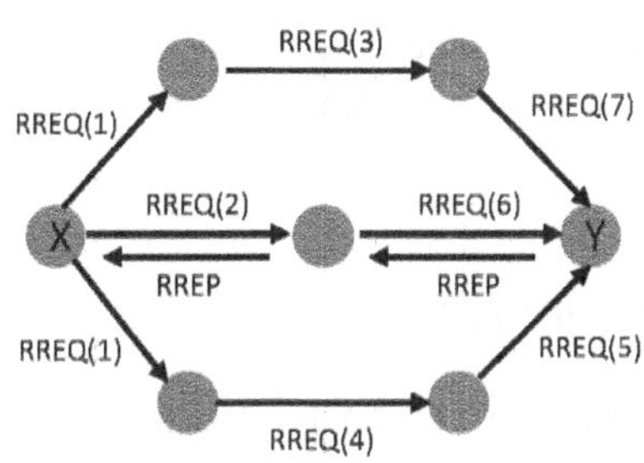

Figure 12.5. *Route discovery flooding*

When a mobile node wishes to send packets to a destination, it starts checking if it has an available route in its routing table. If this is not the case, it will initiate a route discovery by sending a Route Request packet (RREQ) to locate the mobile recipient. This packet is distributed to its neighbors. The mobile receiving this RREQ package looks if they have a route available to the recipient in their routing table. If this is not the case, they in turn rebroadcast the RREQ packet to their neighbors and keep a trace. By using sequence numbers on packets, this allows us to avoid an infinite loop. RREQ packets are flooded throughout the network. When such a packet reaches the destination or a mobile node that has a route to this recipient, then a response packet (RREP) is generated by the mobile and sent back again, due to information stored in the caches traversed by the mobile RREQ. During the flooding of the RREP packet, each mobile node will update its routing table by keeping the identity of the destination according to the route selected by the latter. For example, in Figure 12.5, the mobile node X, which has no route to mobile node Y, initiates a route discovery process. As we could see, the RREQ generates many redundancies and uses a high number of resources. From the MAC layer point of view, many nodes will compete in order to win a slot within the conflict resolution protocol. When the number of requests is very high, this protocol provides a divergence that inhibits completely the network operations. In a dense network (high number of neighbors per node), the flooding is a disaster for the network performance.

AODV today has many successes and is used in many systems. The Zigbee and the WiFi mesh materials use an adapted version of AODV.

12.3.3. *Dynamic source routing*

Dynamic source routing (DSR) is also a reactive protocol and it works in a very similar to AODV. The main difference resides in the routing procedure. AODV is a table-driven routing and DSR is source routing.

Nodes in DSR are idle and when a node decides to send traffic to destination, it will initiate a route discovery in an identical procedure than AODV. Then to indicate the route, instead of creating local routing tables, each intermediate node prints its identity of the node in the packet header. The drawback of source routing is of course the additional overhead in the packets.

The advantage of source routing is that the route information is retained in the packets and nodes could at any moment be aware of the route from the source to the destination. This is not the case with routing tables because the only information that we keep is the next hop. Knowing the complete route could facilitate the reaction of mechanisms for quality of service, security, energy, etc.

12.4. Proactive protocols

A proactive routing protocol maintains regularly updated information on the network at each node. This information is obtained through messages sent periodically or triggered in response to important events occurring in the network. This information allows each mobile to keep topology knowledge offering the possibility to calculate a route for each destination in the network. The routing policy adopted in packet radio networks had a proactive approach. Examples of proactive routing protocols are Destination-Sequenced Distance-Vector (DSDV), Optimized LinkState Routing (OLSR) and Topology Broadcast Based on Reverse-Path Forwarding (TBRPF). OLSR became TBRPF Experimental RFC. DSDV is, however, now completely obsolete.

12.4.1. *Optimized link state routing protocol*

Optimized link state routing protocol (OLSR) aims to maximize the dissemination of information packets in the network and reduce the overhead due to the flooding procedure. For this, the protocol is based on the concept of multipoint relays. A multipont relay (MPR) is authorized to retransmit information to another relay. Each node selects its MPR according to the following rule: from my direct neighbors, I will choose a minimal subset that could cover all two hop neighbors with symmetric links. Figure 12.6 displays a comparison between classical flooding and an MPR optimized one. The more dense the network, the more efficient the MPR optimization. The MPR selection is related to the dominating set problem which is NP-complete and it is necessary to use heuristics to reduce the time needed for the selection.

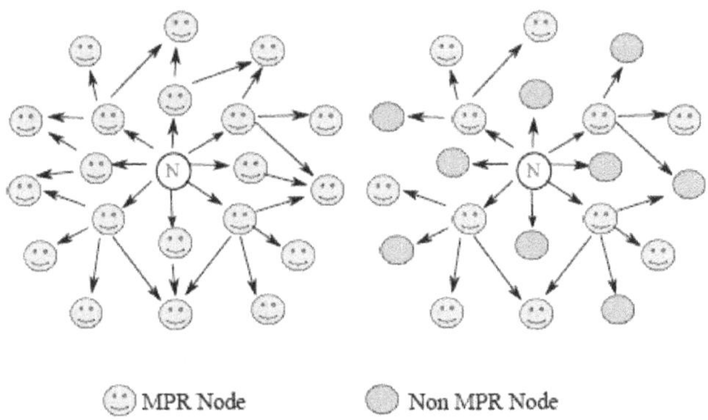

Figure 12.6. *Route discovery flooding*

To establish its list of multipoint relays, each mobile regularly sends its identity and its list of neighbors. With this information, each mobile will know the network topology in its tow-hop neighborhood. Thus, it will be able to determine a list of multipoint relays which will cover all the neighbors in a two-hop distance. Each terminal informs its multipoint relay of its choice, which allows each mobile to know whether it is able to retransmit the information it receives from a given mobile. The second step consists of sending the list of selected MPRs to all the nodes of the network (MPR flooding) and thus each node which receives this information will be able to build a topology map and calculate the best route for each destination.

12.4.2. *Topology dissemination based on reverse-path forwarding*

As with OLSR, topology dissemination based on reverse-path forwarding (TBRPF) is a proactive protocol that seeks to reduce the use of bandwidth. However, and unlike OLSR, which maintains a partial topology of the network (using MPR), all mobiles in TBRPF have a knowledge of the complete topology of the network. Such knowledge is more expensive but allows the use of multiple paths or paths with a certain Quality of Service. For the network topology, each node will maintain a tree of shortest path (it may be possible to use a different metric) to all other nodes. The tree is unique to each mobile. When a mobile detects a state change at one of its neighboring links, it sends an update on its associated tree. Each internal node in the tree forwards the update message.

The trees themselves are constructed according to the update messages: when a node receives such a message, it computes the tree-related message initiator. The update messages are sent only when detecting a status change (appearance or disappearance of a link). Only the neighbor discovery messages are sent periodically.

12.5. Quality of Service in ad hoc networks

For the last several years, there has been a significant increase in interest in supporting Quality of Service (QoS) constraints in MANETs. The specificity of MANETs makes existing solutions for wire line networks not very suitable and a broad range of novel approaches have been studied. When considering QoS support, there are several aspects that must be taken care of.

First, one must choose a QoS model, i.e. the eventual goals one wants to achieve. These are closely related to the applications that are to take advantage of the QoS support. Real-time audio/video applications require constant bandwidth and little delay, whereas FTP transfers can go with no more than classical best-effort. Moreover, the former can adapt its throughput to the available bandwidth and sustain data loss to some extent, while the latter requires a lossless channel. Most of the time, the model must comprise bandwidth and delay constraints. Second, a QoS-aware routing protocol should be used in order to find a suitable multi-hop path given some link constraints supported by the model. Third, if reservation is to be made – which is the case for most QoS-requiring applications, though a strict reservation is possibly not suitable, because of the dynamic topology variations – a signaling system has to be chosen to communicate between the source node and all the intermediates. Likewise, in the case of constraint breakage, the signaling is used by the intermediate nodes either to repair the link – by reestablishing a bypassing path – or to report the failure to the source node. Finally, when the connection ends, the reservation has to be torn down. Once more, because of the varying topology, the reservation is maintained in a soft state, i.e. it is never definitive and has to be renewed on a regular basis. This is due to the fact that a link failure may cause a partition of the network that would not allow us to contact intermediates anymore. Signaling generally comes in two flavors: in-band and out-of-band. The first consists of control information piggybacked on regular data packets (e.g. as options in the IP header) and has the advantage of not transmitting additional packets that would contend with data packets and potentially waste bandwidth or delay. The second is a special kind of control packet that can be transmitted independently of any data flow.

Finally, nodes must support some packet scheduling (fair queuing) algorithms in order to prioritize some packets or to shape some flows. Optionally, if the MAC layer provides some kind of support for QoS, it can be taken advantage of in upper layers. Support can exist in direct low level delay respect or bandwidth allocation (e.g. in TDMA-based MAC protocols) or solely in link quality accounting capability.

12.6. Models for QoS in MANET

In this section, we present works carried out to provide QoS issues in MANET:

– Flexible QoS Model for MANETs (FQMM) is an attempt to build a model explicitly for MANETs. It can be seen as a mix of IntServ and DiffServ in the way that it supports both per-flow and per-class granularity. Flows with borderline QoS requirements are granted per-flow processing, while the others are aggregated in classes. As in DiffServ, there are three types of nodes: the sender (ingress node), the routers (interior nodes) and the receiver (egress node). The ingress node is required to police its outgoing traffic to meet a given traffic profile. Traffic profiles are expressed in terms of percentage of available bandwidth, to cope with changing link conditions. The routing algorithm is left unspecified and is assumed to be of multi-path kinds (i.e. all routes to a target are known allowing for multiple choices with respect to some bandwidth requirements), though the use of some QoS-aware routing protocol is encouraged.

– Core-Extraction Distributed Ad hoc Routing (CEDAR) is a reactive protocol which optimizes routing request by using core nodes, i.e. nodes that belong to the dominating set of the network. Each node broadcasts periodically BEACON messages which are used for one-hop neighborhood discovery. In a first phase of the protocol, each node must elect its core node. The set of nodes that are the core node of a neighbor is called the core of the network. Then, local topology information is disseminated by means of piggybacking link information to BEACON messages to the three-hop neighborhood of the nodes, allowing each core node to maintain virtual links to its neighboring core nodes. Thus, core nodes maintain information about the local core graph. When a node s needs a route to d, it requests a route calculation to its core node, dom(s). By means of some chosen reactive routing protocol, dom(s) finds a core path to dom(d). After having learnt the route to dom(d), s sends packets to d by means of source-routing. Now, if QoS constraints are to be honored for a given route calculation, each core node must find a virtual link to the next core node toward dom(d) that satisfies them. The use of core nodes for route calculation assumes that the core path is a good hint toward the optimal route. But, it does not ensure that the best route can be found.

– Ticket-based probing aims to achieve a near-optimal performance while avoiding flooding and accounting for link information imprecision. QoS satisfying routes are found by sending probes carrying some limited total amount of tickets toward the destination using distance vector information. At each intermediate node, a probe may be divided into several ones with the tickets distributed between them. The total amount of tickets is to remain the same along the propagation. The goal is to maintain a compromise between the odds of finding an optimal route (more tickets sent) and reduced control overhead (less tickets sent). If only one ticket is sent, then the plain old shortest path is searched. TBP relies on the assumption that stable links tend to remain stable contrary to the so-called transient links. Each node i collects statistical information about delay Di(t), bandwidth Bi(t) and cost Ci(t) to each other node t in the network. It is thus not meant to be scalable, but is instead destined for moderate scale networks. Along with Di(t) and Bi(t), each node maintains the association variation DDi(t) and DBi(t) by which the next reported value will differ from the current one. The protocol makes a distinction between delay and bandwidth constrained paths. Thus, it either searches a route for least-cost delay-constrained connections, or least-cost bandwidth-constrained connections. For each purpose, it defines two kinds of tickets – yellow ones and green ones – that make the probes either look for feasible paths or least-cost paths. The relative quantity of each kind governs which kind of path is going to be searched.

– INSIGNIA is meant to provide a complete framework for QoS support in MANET. It is based on the assumption that applications requiring QoS support also provide a way to modulate this requirement and can sustain service degradation by the use of adaptive services. The model thus defines three operating modes for a flow: best effort (BE), base QoS (BQ) constraints with minimum bandwidth and enhanced QoS (EQ) with maximum bandwidth. It also assumes, quite rightfully, that the destination node is the best place to measure the QoS of a flow and make decisions about service degradation. The framework provides a form of in-band signaling between source and destination by the addition of an option in the IP header of packets, the purpose of which is to attach QoS requirements to a flow as well as a means to alert the destination in case the requirements are not fulfilled at a bottleneck node.

The framework is not specifically designed to operate with a given routing protocol and is instead meant to support any MANET routing protocol. Thus, it does not take any advantage of QoS-aware routing but rather relies on service adaptation on the applications side. Moreover, the framework remains independent of any MAC layer specificities which should ensure easy implementation with existing networking equipment. The destination is required to report the QoS level to the source in the feedback in order to modulate the flow to suit QoS conditions. Rerouting is provided by the routing protocol and quality is subject to degradation in case of failure to reserve resources along the new path. Old reservations are nevertheless supported by soft state and disappear if not regularly refreshed.

Stateless Wireless Ad hoc Network (SWAN) provides QoS support through service differentiation. It is another framework independent of the routing protocol and the MAC layer, though it relies on the latter for bandwidth and delay measurement. The novelty is that routing nodes do not maintain any state information about the flows they forward – hence they are stateless. Each node differentiates two kinds of traffic through DSCP flags in the IP header: real time (RT) and best effort (BE). RT flows are forwarded normally, whereas BE are fed to a traffic conditioner. The node actively monitors RT bandwidth usage and sets the conditioners' parameters accordingly to shape BE traffic to the residual bandwidth so that packet queues are maintained in short form to avoid delay increase. When a node intends to open a QoS constrained connection, it sends a probe to measure bottleneck bandwidth and available delay. Here, it seems that any routing protocol may be used, especially a QoS-aware one comes in handy. While traversing the network toward the destination, the probe is marked by the routing nodes with the measure actual link delay and minimum bandwidth. The destination copies this gathered information in the reply packet and sends it back to the source which can then decide whether to admit the flow or not. Once the flow is admitted, it has to be routed through the chosen path.

Quality of Service for OLSR (QOLSR) is an extension of OLSR protocol which is capable of providing multiple-metric routing. This extension can be added to OLSR functioning. No additional control traffic is generated (only augmented HELLO and TC messages). QOLSR protocol uses standard multipoint relays (MPRs) to ensure that the overhead is as low as possible for forwarding control traffic. Local QoS metrics information on links is used to elect the Quality of Service MPRs (QMPRS). This information is then flooded in the network by TC messages to calculate routing tables. QOLSR can find optimal paths on the known partial topology having the same bandwidth performances as those on the whole network. These paths contain only QMPRs as intermediate nodes between a source destination pair.

QOLSR carries out different functions which are required to perform the task of routing: (1) neighbor sensing allows, by means of hellos messages, each node to detect the neighbor nodes with which it has a direct and bi-directional lin; (2) neighbor QoS measurement is done by the node to estimate the QoS conditions (available bandwidth, delay, loss probability, cost, security, power consumption, etc.) on links to each neighbor having direct and symmetric link; (3) Quality of Service multipoint relay selection which is made by the node to design its own set of QMPRs. This set is calculated to contain a subset of the 1-hop neighbors which provides maximum bandwidth and minimum delay from each 2-hop neighbor to the given node. The QMPR set does not need to be optimal, but it should be small enough to minimize the number of generated TC messages in the network. Topology Control extension messages are sent to all the nodes by each QMPR node regular interval to declare its QMPR selector set and QoS conditions; and (4) routing table calculation: each node maintains a routing table which allows it to route packets for the other destinations in the network with optimal metrics respecting QoS constraints.

12.7. Mesh networks

Mesh networks are networks of access points forming a mesh together. Customers are connected on access points and the access points are connected by wireless links forming a network.

The advantage of these systems is that they can cover a large geographical area without the need for cables. For example, on a large campus, the access points can be set on the roofs of various buildings thus avoiding the use of wired Ethernet. Several opportunities are emerging to make a mesh network:

– use the same frequency at all the terminals and the mesh nodes considering that the access points are treated at the same time as routers. The difficulty is to efficiently use the bandwidths that are shared by terminal machines and nodes. Two access points should not be too far away and should not require us at to transmit and receive in the same time to avoid lowering their speed. This solution at considered as the first-generation mesh networks;

– using different frequencies for terminal equipment and mesh nodes. For example, using a 802.11b/gWiFi network with the three frequencies available makes it possible to use two communication cards with different frequencies. The downside is of course disturbing the frequency plan, especially if the network is large and has many access points. This solution is part of the second-generation mesh networks;

– still in the second generation, the mesh network uses a different standard to connect the access points to one another. For example, a mesh network that connects guests with 802.11g and 802.11a is used for interconnecting the access points;

– the third generation uses three frequencies. One to connect customers and the other to interconnect the two access points. In this case, the upstream and downstream connections of the same node use different frequencies. It usually uses 802.11a, which has up to eight different frequencies.

The mesh networks raise new problems for wireless networks, including: how to optimize battery of the access points if they are using battery due, for example, to the mobility of the nodes? How to optimize routing to avoid disturbing the user traffic from access points, especially if they are already saturated? What density of access points should be used, which is to question the power of the access points?

The advantage of this technology is to be able to easily reconfigure when an access point fails. Guests can connect to another access point, even slightly increase the power of neighboring access points that failed.

The main problem is in managing the routing scheme. This routing scheme is treated in the access points that are not very powerful machines, and must therefore avoid access points that support a lot of traffic from connected customers. Many proposals have been made, primarily those derived from ad hoc networks.

Following 15 proposals, the working group adopted two proposals IEEE 802.11sa the SEE-Mesh and Wi-Mesh, which have come together to form a single proposal. This proposal became a standard in April 2007, after many discussions about implementation. The access points and stations that have the 802.11s routing algorithm are appointed Mesh Points (MP). Radio links are used to interconnect them. The default protocol is Hybrid Wireless Mesh Protocol (HWMP). This hybrid protocol is a combination of a protocol from the AODV, the RM-AODV (AODV Radio-Metric) and an algorithm based on the trees. A second protocol may be used when MP is accepted: RA-OLSR (Radio-Aware OLSR).

The IEEE 802.11s group also defines security solutions for mesh networks. To do this, it defines mutual MP authentication, generates and controls the session keys, enables data encryption on the lines of the ad hoc network and detects attacks. For this, it is necessary to perform authentication with the IEEE 802.1x protocol. The session keys are managed by a PKI (Public Key Infrastructure). Confidentiality is ensured by the IEEE 802.11i standard.

Handovers are possible in some mesh networks, allowing a greater mobility for the mobile equipment. This is the case of the Green Communications mesh network that will be studied later on. Moreover, telephony is becoming a special application of these environments. We can therefore expect a diversification of handheld terminal stations able to connect to networks of ambient Internet available in all popular crossing points, such as the centers of cities, railway stations, airports, subways, etc.

The advantage of ad hoc networks and mesh over 3G/4G technologies is to use the frequencies over very limited areas, allowing a high reuse. In addition, the cost of a mesh access point is negligible compared to that of a 3G/4G antenna. Given the explosion of mobile flows, the gain realized by an ad hoc or a mesh network is enormous. Most working groups dedicated to mobile networks are looking at hybrid solutions to save both energy and spectrum.

Another advantage of mesh networks over 3G/4G comes from energy consumption. We address this issue in more detail in the following section, but in short, a 3G/4G consumes on the average 2,000 W per antenna. Concerning mesh access points, they use less than 20 W. Substituting a 3G/4G cell by a set of 30 mesh points, the overall throughput of the mesh network is much higher, and the consumption is halved. We will see later on how to decrease the energy consumption of mesh access points when the traffic is low, such as at night.

12.8. VANET networks

VANET is a network concept from the so-called Intelligent Transportation Systems. As shown in Figure 12.7, vehicles communicate not only with each other by intervehicle bonds, or InterVehicle Communication, but also with facilities located along the road via vehicle communication equipment or Roadside-to-Vehicle Communication. These vehicular networks contribute to safer and more efficient roads in the future by providing timely information to drivers and the authorities concerned.

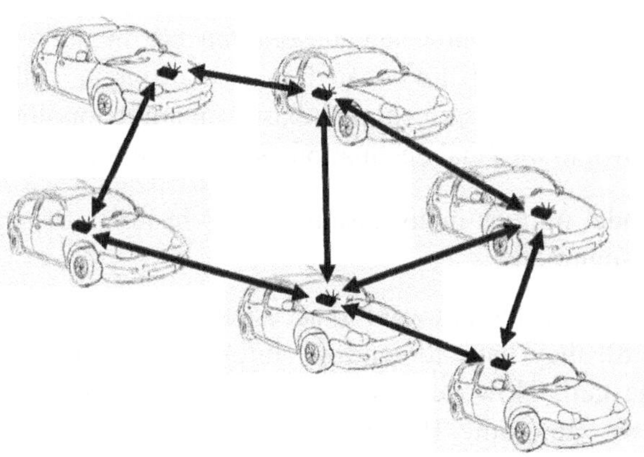

Figure 12.7. *A VANET network*

Many applications have been developed as part of the road network, such as reporting a problem on a road. A car accident can automatically send an alert signal to the vehicles that are following or that will shortly arrive at the accident. Green lights can be synchronized with the approach of a car. Self-diagnosis of engine may, after connecting to a cloud, send a warning to the driver indicating the problem and the associated diagnosis. The VANET networks are quite similar in this case to MANET networks, but with the distinction of having fixed customers, such as traffic lights and signs. A haracteristic of VANET networks, compared with MANET networks, is the greatest simplicity of travel related to roads and railways, which generally determines the direction taken by the customer.

Another case is represented by a vehicle carrying many passengers. Many applications have been developed for these types of transport, such as aircraft multiplayer games, or discussions between people who do not know one another, but who will be together for several hours, etc.

12.9. Green PI: wearable Device2Device networks

Telecommunication technologies in all their forms (2G to 4G, WiFi, WiMAX, etc.) use a unique model to communicate between two entities. It consists of forwarding the packets to an access point which sends the information back to the destination (Figure 12.8). This happens when source and destination are in the same cell. If source and destination are located in two different cells, then the information uses a hierarchical path based on a tree topology going up and then down to the final destination as shown in Figure 12.8.

This model implies many disadvantages in terms of capacity and energy consumption:

1) When many devices want to access the network in the same cell (Figure 12.8), they have to send the information to the access point serving the cell. However, the access point can only process one request at the same time. If many requests arrive simultaneously, they generate collisions and interferences. A scheduling mechanism needs to be elaborated at the access point in order to organize the access to

the network, request by request. Parallel request cannot be served which limits the capacity of the cell.

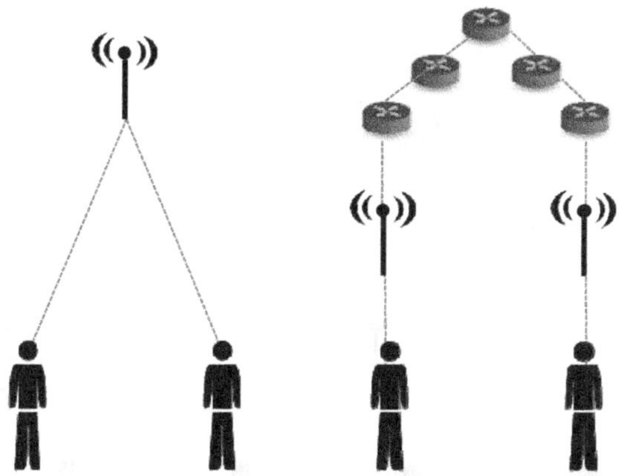

a) b)

Figure 12.8. *a) Communication in one cell and b) in different cells*

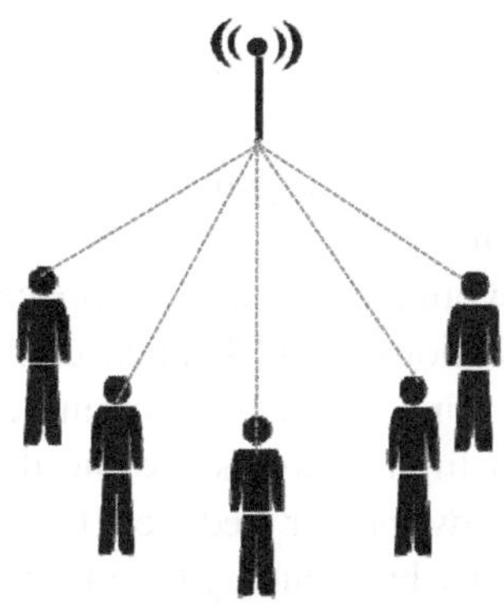

Figure 12.9. *Scheduling at the access point*

2) In terms of energy consumption, transferring the information to an intermediate access point instead of direct transfer doubles the number of transmissions/receptions. Also, when using a tree-topology path to climb up then down on a vertical way will increase the energy per bit comparing to a direct and horizontal path. Figure 12.10

illustrates the difference between horizontal and vertical communications.

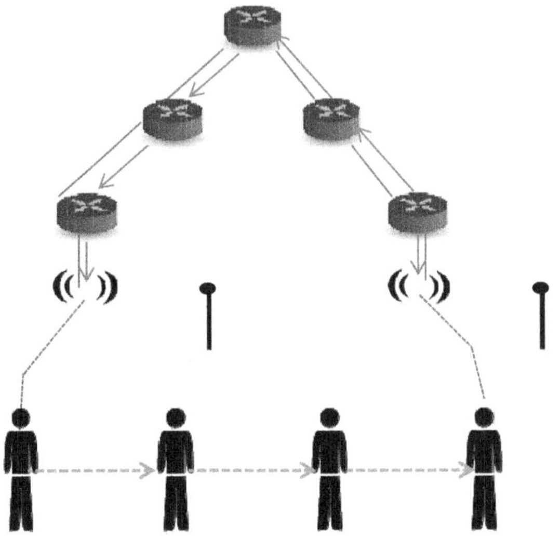

Figure 12.10. *Horizontal versus vertical communications*

The objectives of the next generation 5G is to increase the network capacity by introducing a Device2Device technology. Using direct communications would increase the capacity of the networks since parallel communications could happen comparing to an access point centralized scheduling. Direct communications do not need to communicate with high power since the destination is very close to the source. The low power reduces the energy consumption and increases the capacity by reducing the interferences.

12.9.1. *Observation of traffic*

When scanning the traffic on telecommunication networks, one could observe that an important part of the traffic is local. People during events or in smart cities and smart homes share pictures and videos, exchange information on their environment. The information in the majority of cases does not need to travel through the infrastructure of the network and the Cloud and to consume a huge amount of resources and energy to return to the neighborhood. However, as the video traffic on mobile devices is doubling every year, this avoids to create bottlenecks at the access points.

12.9.2. *Embedded Internet and 5G*

The idea to go directly from source to destination is not new. Its implementation has never been realized in the past because operators of networks had to control the traffic for accounting and monitoring. Nowadays, the majority of subscriptions include an unlimited usage of the network. Hence, the more the traffic is out of the operator network, the less the infrastructure of the operator is saturated while its income stays the same. Introducing horizontal and direct communications is a win-win policy for operators, customers and also the entire eco-system. Why, should local traffic travel over the entire network and data centers just to reach a close destination?

5G networks include in their requirements the faculty that two devices should communicate directly with no need for any network infrastructure. This implies that the device should contain a minimum of network intelligence to have the capacity to detect destinations and to also transfer the traffic of its peers. In addition, Device2Device communication should continue to work even when the global network is not operating or when damages push the global network offline.

Attempt to create intelligence in the applications of Smartphones is an option to introduce the Device2Device. Some products are already working on different platforms (IOS, Android, etc.) such as AirDrop or Firechat. In those cases, the Device2Device is managed at the application layer and with a proprietary protocol. This means that Smartphones should first install the application to run a Device2Device. Furthermore, each new service requires the development and download of a new application. This model cannot scale.

12.9.3. *Green PI: wearable and embedded Internet*

Green Communications has introduced the concept of Embedded Internet where devices hold a full TCP/IP environment, an open operating system (Buildroot-based Linux) and a sharing storage. Each device contains two wireless network interfaces. One to establish the network between the devices and the second to offer access to SmartPhones, PADS and computers. All types of applications will be active immediately because the network is running a TCP/IP locally. It is also possible to provide many applications over http using a local web server and the latest web technologies (htmls). This gives the opportunity to anyone to communicate even if they have not downloaded a native application.

The Green Communications' device is light enough to be held or wearable. It could be distributed on-site or could be brought along by the visitors. When turning the devices ON, the network is established in a few seconds and then a local Internet is operated. A subset of the devices could be configured as gateways to the global Internet to create a hybrid model where local traffic stays local and global traffic is transferred across the gateways.

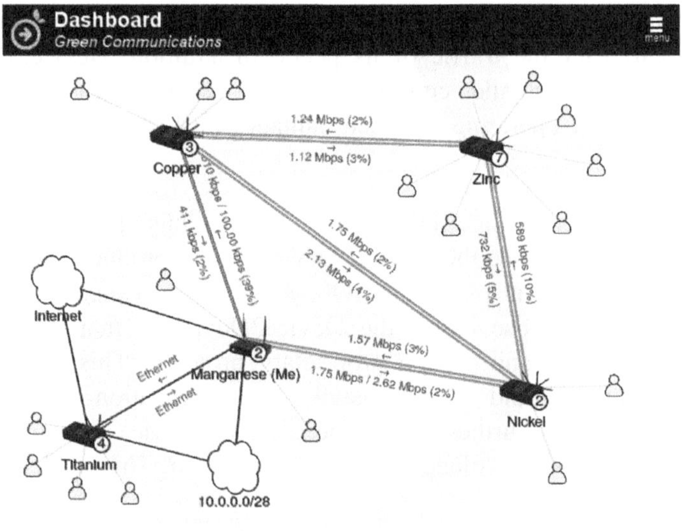

Figure 12.11. *Monitoring interface with open data*

The device also includes storage with security options to offer the possibility of sharing information and also installing new specific applications related to the area where the network is running. The network provides access to its raw open data to help people in their connections and enables application developers to build more appropriate applications related to on-site usage. Figure 12.11 shows the different kinds of data that are open: network topology, number of users per access point, quality of links between access points, location of the Internet Gateways, access point profiles through the vCard Standard, etc.

The idea that people could hold their own personal boxes creates the concept of the Participatory Internet (PI). Any group of persons could simply, by turning ON their own boxes, create an Internet that could operate on local content or connect to the global Internet.

Green PI offers advantages to facilitate the communication in many cases:

1) Network densification: dense areas such as stadiums and large areas need a strong network densification to offer Internet access with acceptable quality. Providing an access point for every 30 users is cash intensive and needs heavy civil engineering schemes for cabling. Asking a subset of participants to hold a Green PI access point reduces the cost of the coverage by a factor of seven.

2) Mobile units: connecting mobile units (robots, drones and people on the move) is not possible with an infrastructure-based network (4G, satellite, etc.). A participatory Internet with an embedded TCP/IP and local content is the solution to create an on demand network and allow direct communications.

12.9.4. *Distributed TCP/IP*

In order for an embedded Internet to work correctly, it is mandatory to distribute all the services over the devices. Services should not depend on one specific machine to avoid any network blackout. When a device enters or leaves the network, services must continue to work correctly. Green Communications provides distributed TCP/IP over the devices by enhancing them with Zeroconf (mDNS, DNS-SD, etc.), anycast addresses, handoff management, distributed DHCP implementation, distributed database, etc.

12.9.5. *Wearable YOI*

The Wearable YOI consists of hardware containing three network interfaces (two wireless plus Ethernet) and working with a Buildroot-based Linux. It offers embedded TCP/IP providing WiFi access to the local or global Internet. All native TCP/IP applications may be active in private or public connections.

The performances of Green Communications products have proved to provide voice over IP and video conferencing over 10 WiFi hops. This unique performance has never been shown before by any competitor.

Here is a subset of the Wearable YOI specifications:

Nano YOI	Wearable YOI (coming soon)
Raspberry Pi B, B+ or A type	Customized board
Weight 200 g	Weight 300 g
-2 WiFi USB interfaces Wi–Fi 802.11n MIMO 2 × 2	2 MiniPCIe WiFi with internal antennas
-	Embedded Battery 10500 mAh
-	XMPP local distributed server
-	Distributed DHCP for the access
VLAN on access and backhaul	
Wireless link capacity estimation	
Routing according to the highest capacity	
Capacity reservation per WiFi link	
Unused capacity redistribution between WiFi links	
Network open data	
WPA2 Enterprise on network access and network backhaul	
Embedded TCP/IP, Zeroconf compliant	
Storage capacity	
Start and Stop mechanism for unused devices	
Auto-configuration for the backhaul, handoff, plug&play	

12.10. Bibligraphy

[AGG 04] AGGELOU G., *Mobile Ad Hoc Networks: From Wireless LANs to 4G Networks*, McGraw-Hill Professional, 1st edition, 2004.

[BRO 95] BRODSKY I., *Wireless: The Revolution in Personal Telecommunications*, Artech House, 1995.

[BUN 00] BUN LEE S., AHN G.S., ZHANG X. *et al.*, “Insignia: An ip-based quality of service framework for mobile ad hoc networks”, *Journal of Parallel and Distributed Computing*, vol. 60, pp. 374–406, 2000.

[BUR 02] BURKHART J., *Pervasive Computing: Technology and Architecture of Mobile Internet Applications*, Addison Wesley, 2002.

[CHE 99] CHEN S., NAHRSTEDT K., "Distributed quality-of-service routing in ad-hoc networks”, *IEEE Journal on Selected Areas in Communications*, vol. 17, pp. 1488–1505, 1999.

[CLA 03] CLAUSEN T., JACQUET P., Optimized link state routing (OLSR) protocol, RFC 3626, IETF, 2003.

[JOH 07] JOHNSON D., MALTZ D., The dynamic source routing protocol (DSR), RFC 4728, IETF, 2007.

[JUB 87] JUBIN J., “The darpa packet radio network protocols”, *Proceedings of the IEEE*, vol. 75, pp. p. 21–32, 1987.

[OGI 04] OGIER R., TEMPLIN F., LEWIS M., Topology dissemination based on reverse-path forwarding (TBRPF), RFC 3684, IETF, 2004.

[PER 94] PERKINS C., BHAGWAT P., “Highly dynamic destination sequenced distance-vector routing (dsdv) for mobile computers”, *ACM SIGCOMM'94*, 1994.

[PER 03] PERKINS C., BELDING-ROYER E., DAS S., Ad hoc on-demand distance vector (AODV) routing, RFC 3561, IETF, 2003.

[PRA 98] PRASAD R., *Universal Wireless Personal Communications*, Artech House, 1998.

[SIV 99] SIVAKUMAR R., SINHA P., BHARGHAVAN V., "Cedar: a core-extraction distributed ad hoc routing algorithm”, *IEEE Journal on Selected Areas in Communications,* vol. 17, pp. 1454–1465, 1999.

[TOH 94] TOH C.K., *Ad Hoc Mobile Wireless Networks: Protocols and Systems*, Prentice Hall, 2001.

[TSE 05] TSE D., VISWANATH P., *Fundamentals of Wireless Communication*, Cambridge University Press, 2005.

[XIA 00] XIAO H., SEAH W.K.G., LO A., "A flexible quality of service model for mobile ad-hoc networks", *IEEE VTC*, vol. 1, pp. 445–449, 2000.

13

Mobile-Edge Computing

13.1. Network virtualization

Virtualization is not a new technique, since it was introduced in the first mainframe computers to share the memory through the paging system. The trick was to bring back the memory pages from HDD RAM just before the CPU needs them. Thereafter, a number of mechanisms have been implemented with this solution so that the user has the impression that the services needed are located on a nearby machine when they actually are on a remote machine.

Server virtualization has been a tremendous success for grouping multiple servers on a single machine; the user has the impression of yet having only one server. Network virtualization holds the same principle: several virtual networks share the same physical infrastructure.

Network virtualization was initiated by an American project called Global Environment for Network Innovations, in which Intel intended to build a router consisting only of its material part, without any network operating system. Software companies had only to implement their system on the machine. The advantage of virtualization lies precisely in the ability to implement multiple network operating systems on a single machine using a "hypervisor". A virtual router is shown in Figure 13.1.

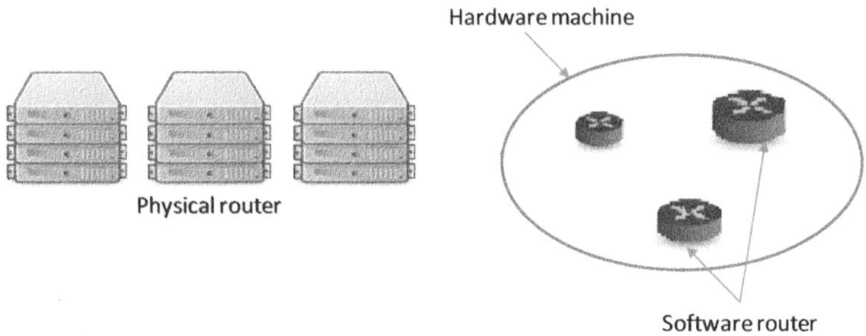

Figure 13.1. *A virtual router*

A virtual router is therefore a logical instance of a physical router and several virtual routers can run simultaneously on one physical router, or on a generic machine with input-output required to achieve the physical router. We then obtain the ability to deploy multiple networks that can be called virtual networks on a single physical network. A virtual network is a grouping of a set of compatible virtual routers. Virtual routers use the physical links to interconnect.

Such an environment is shown in Figure 13.2, in which the networks A, B, C and D coexist on the same router and the same physical lines.

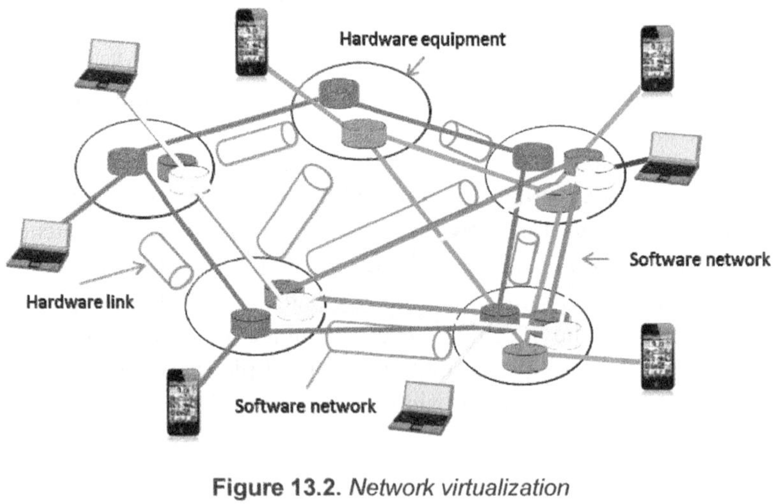

Figure 13.2. *Network virtualization*

A virtual network is made from the virtual routers with the operating system A. Similarly, the virtual network B is formed with the bodies B and so on. Each instance of the same physical router can use completely different protocols, networks are also totally different. Everyone can use the protocols related to a data application.

The advantages of this virtualization technique are numerous. The first is to be able to run on the same physical network, virtual networks using completely different technologies. One can, for example, get a first network using the IOS network operating system from Cisco, a second using JUNOS Juniper, a third and a fourth using the Nokia and Ericsson systems, etc. It is also possible to run several different releases of the same network operating system. Obviously, the above examples have little chance of being realized because network manufacturers do not necessarily want to market their software without their equipment. Decoupling hardware/software will only be available for open systems.

A second advantage of virtualization is the use of resources. It is obvious that the resources can be much better utilized by starting and stopping virtual routers or by moving an overloaded virtual router. It is possible to move one or more virtual routers to other physical routers.

In all cases, the resources must be completely isolated. This implies a great security between networks. If any of the virtual routers fails, this does not affect the others. If, of course, a physical routers fails, all virtual routers will also be down. The isolation also protects against network attacks.

13.2. Network virtualization technology

Both technologies can be used for network virtualization. Xen is one of the solutions for virtualization among many others but certainly one of the oldest solutions. OpenFlow is a signaling system for virtualized networks of the new generation: Software-Defined Networking (SDN). We are going to describe these two environments.

13.2.1. *Xen*

Xen is a Virtual Machine Monitor (VMM), open source, also called a hypervisor, that runs on standard hardware platforms. In addition to the VMM, on the physical hardware, the Xen architecture is composed of several domains running simultaneously on the hypervisor, called virtual machines (see Figure 13.3). Each virtual machine can have its own operating system and applications. The VMM controls access to multiple hardware domains and manages the sharing of resources between different areas. Thus, one of the main tasks of VMM is to isolate different virtual machines, so that the execution of a virtual machine does not affect the performance of other machines.

All device drivers are stored in an isolated area that is reserved for them. Called domain 0 (dom0), it provides them with a reliable and efficient hardware support. Domain 0 has special privileges compared to other domains, called user domains (domU), and has, for example, full access to the hardware of the physical machine. The user domains have virtual drivers and operate as if they could directly access the hardware. However, these virtual pilots communicate with the dom0 in order to have access to the physical hardware.

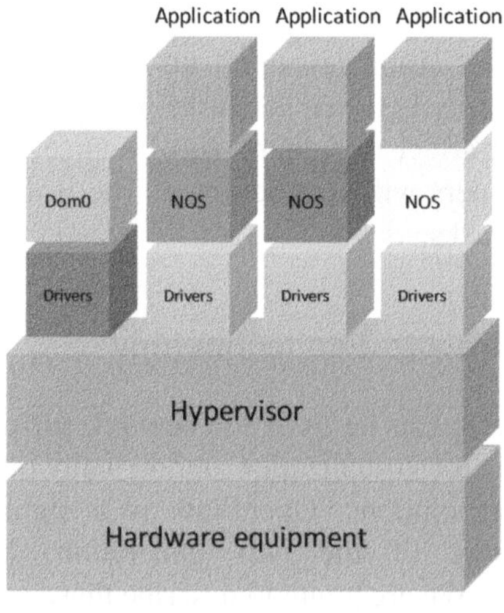

Figure 13.3. *Hypervisor and virtual machines*

Xen virtualizes a single physical network interface inbound packet by demultiplexing of the physical interface to the user domains and, inversely, by multiplexing the outgoing packets generated by these user domains. In this procedure, called virtualization of networks, the area 0 directly accesses the peripheral IO using its native device drivers and performs input-output operations from the domU.

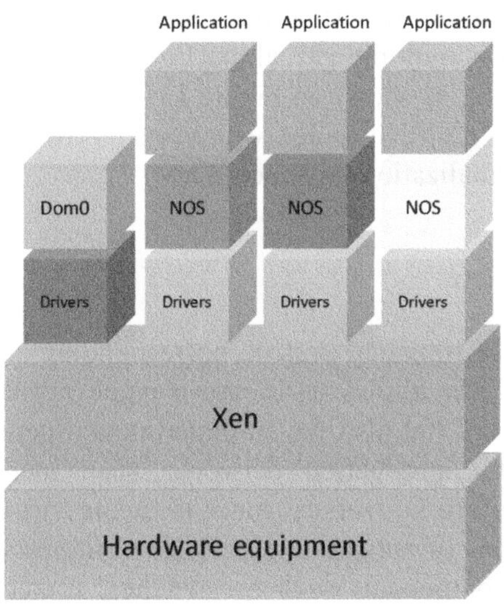

Figure 13.4. *Virtualization with the Xen hypervisor*

The user domains employ virtual input-output devices controlled by virtual drivers to ask the dom0 for access to the device, as shown in Figure 13.4. Each user domain has its own virtual network interfaces, called foreground interfaces, required for network communications. The background interfaces are created in the area 0 corresponding to each leading interface in a user domain and act as a proxy for the virtual interfaces in dom0. Leading interfaces and background are connected to each other via a channel input-output, which employs a zero-copy mechanism to match the physical page containing the package and the target domain. In this way, the packets are exchanged between the background and foreground interfaces. Leading interfaces are perceived by the operating systems running on user fields as real interfaces. However, the background in the dom0 interfaces is connected to the physical interface and to one another via a virtual network bridge. This is the default architecture, called the bridge mode, used by Xen. Thus, both the input-output channel and the bridge network establish a communications path between the virtual interfaces created in the user domains and the physical interface. Different elements of the virtual network may be implemented using Xen since it allows multiple virtual machines to run simultaneously on the same hardware, as shown in Figure 13.4. In this case, each virtual machine operates a virtual router, which has its own control plans and data, the Xen virtualization layer being low.

13.2.2. *OpenFlow*

OpenFlow is a signaling system, corresponding at the southbound interface. OpenFlow allows us to use a patch not only to set up a production network but also for an experimental network completely isolated from the production network. The OpenFlow interface, initiated by Stanford University, aims to create virtual environments for new categories of networks in parallel to the production network using network components such as switches, routers, access points and PCs.

OpenFlow provides a new architecture to provide virtual network environments. The underlying idea is the physical separation of control and data planes. That is why different elements perform the packet transfer procedure (data plan) and network control procedure (control plane). The transfer is made through a forwarding table that represents the data plane, while the control plane is centralized around a controller. This controller is able to put in correspondence different applications that control each virtual network. An example of running OpenFlow is given in Figure 13.5.

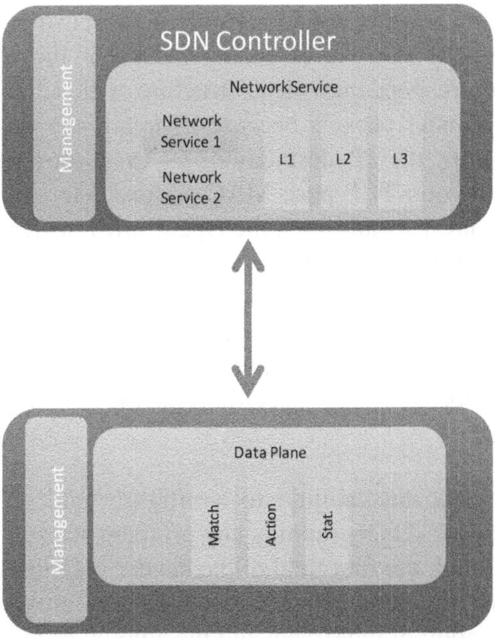

Figure 13.5. *OpenFlow signaling*

The OpenFlow protocol defines the communication between the transfer nodes and the network controller. It establishes a secure channel between each node and the transfer controller using this channel to monitor and configure the transfer nodes. Each time a new packet reaches a transfer node and there is no path configured previously, the first bits of the packet are transmitted to the controller that sets a path for the corresponding stream. The controller may also select the normal processing action for a flow for transmission according to the conventional transfer of the layer 2 an /or layer 3, as if OpenFlow did not exist. This is why OpenFlow can be used in parallel with a production network without affecting the ongoing traffic.

The data plan in OpenFlow is a set of tables described by header fields, meters and actions. The header fields form a structure that describes the 12 positions in the packet header, as shown in Figure 13.6. These fields specify a stream by setting a value for each field or by using a wild card to set only a subset of fields. The flow table also supports the use of subnet masks, if the equipment also supports the correspondence. This structure with 12 subfields leaves flexibility to transfer packets because a stream can be routed on the basis of not only the IP destination, as in a conventional TCP/IP network, but also on TCP port, MAC address, etc. As flows may be determined by layer 2 addresses, the OpenFlow transfer nodes are also called OpenFlow switches. However, this does not necessarily apply and packets can be sent through a layer 3. One of the future goals of OpenFlow is to manipulate header fields defined by the user, which means that the packet header will not be described by fixed fields, but by a combination of fields specified by the administrators of the virtual network.

OpenFlow has the ability to configure nodes to allow the transmission of packets by running any protocol stack. The description of the flow through the first field of the header is followed by counters which are used for monitoring nodes. Counters provide information such as the duration of the stream and the amount of bytes transmitted. The last field in the description of the stream decides on actions. These actions form an instruction set which is applied to each packet of a specific flow in the OpenFlow switches. They include not only the routing of an incoming packet to a destination, but also the modification of header fields such as the ID of the VLAN, or source and destination addresses.

The controller is a central element in the network. It can communicate with all nodes to configure the flow tables. The controller uses a network operating system that provides virtual network management applications and basic functions such as network configuration. Thus, the OpenFlow controller works as an interface between network applications and transfer nodes. It provides the basic functions to access the packets of a flow and monitor node.

OpenFlow works with any controller compatible with the OpenFlow protocol, such as NOX. In this case, each control plane consists of a set of applications running on NOX. This is why a virtual network is defined by its control plan, which is a set of applications running on the controller, as shown in Figure 13.7.

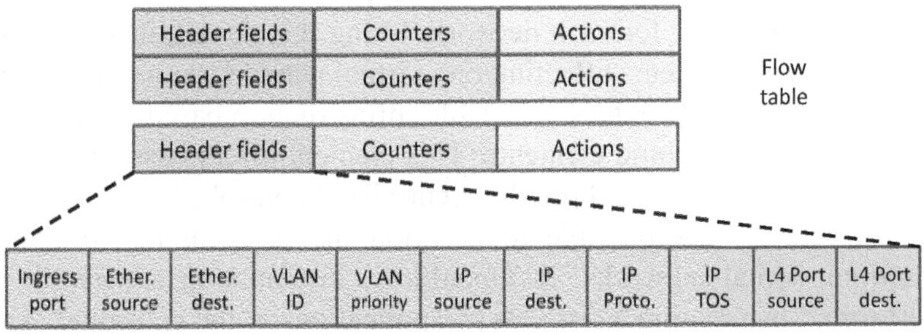

Figure 13.6. *Open Flow meters*

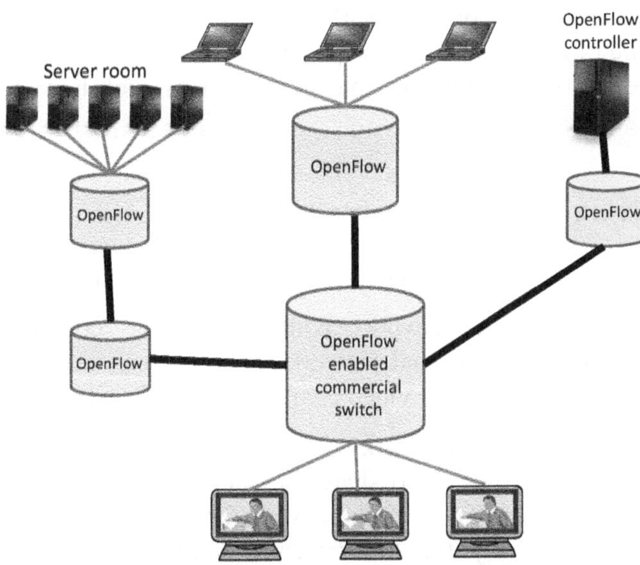

Figure 13.7. *Centralized Open Flow controller model*

By using the single controller model, it is possible to create multiple virtual networks. However, note that the various applications running on the same operating system are in general not well isolated. Accordingly, if a problem occurs in an application, this can stop the operation of the controller, thus affecting all other virtual networks.

FlowVisor is a tool used to allow different OpenFlow controllers to operate on the same physical network. FlowVisor works as a proxy between the switching elements and the controllers, assuming that there is a controller for each network. Using this model, it is possible to ensure that problems affecting one virtual network do not affect the others. OpenFlow provides a scalable infrastructure based on the idea of distributed switching elements. This element offers basic functions for operating a network through a centralized control plan. Using this infrastructure, it is possible to partition the physical network into multiple virtual networks. In OpenFlow, instantiating a network is performed by creating a few sets of applications in the controller. The flows of a new network are accepted on request at the time of the arrival of the first packet entering the network. OpenFlow also provides a flexible infrastructure to reallocate network resources. The reallocation of a network in OpenFlow means only reprogramming the flow table of each participant node. This is a simple operation for the controller which knows the physical devices and how they are connected.

13.3. Using network virtualization

Two types of use can be performed with network virtualization:

– multiple virtual networks within the same company. This solution can create a virtual network for telephony, one for remote monitoring, yet another for video conferencing in the enterprise, etc.;

– several virtual networks of different telecommunication operators. This allows us to share the same physical resources and thus lead to a lower cost for each network. This defines something like a VPN deployed by the infrastructure operator. When a virtual telecommunication operator wants to increase its transmission capacity, this one asks for the infrastructure operator to increase its resources.

Another growing use of virtualization schemes concerns Cloud-based applications. In simple terms, the principle of the Cloud is to put applications as common resources in a Cloud to share these applications with other users. The Cloud offers server resources, computing power, storage capacity, diverse and varied software, etc. Users can share these resources with others, taking into account a strong security to access the information. Virtual resources can migrate from one machine to another one to optimize their utilization. Similarly, the client can move on the Internet and have access to resources from anywhere. To link the resources to the user, it is necessary to create networks. However, these networks can only be virtual because it would be impossible to create a new network as soon as a software migrates or a request is formulated for additional resources.

Finally, virtualization is an ideal technology to test new protocols and new architectures without stopping the operational network. It is quite simple to build a new virtual network using OpenFlow or any other southbound interface. In this case, isolation is an essential property to avoid attacks from one virtual network to another one.

As a summary, virtualization offers many advantages if security can be maintained. Virtualization is seen as an opportunity to provide flexibility to the networks. In addition, this allows us to switch from one technology to another without complicating life too much: we may add virtual networks using new protocols while retaining virtual networks using previous protocols.

The disadvantages of virtualization are linked to the new charge provided by the hypervisor and the need for more memory and computing power, especially if we want to fully separate the different virtual networks.

13.3.1. *Isolation*

The isolation is a fundamental principle of network virtualization: a network cannot have an impact on other networks.

A first simple but very effective solution would be to partition resources between the different networks. In this case, there would not be much to win, because resources of unused network at time t cannot be used by other virtual networks. We must find a solution for both partitioning the resources and ensuring that unused resources can benefit other networks.

This property can be obtained by specific schedulers. The easiest way is to have nodes with as many queues as virtual networks. In each queue, the virtual networks grant credits based on the resources that were allocated to the node for each virtual network. Each network uses a credit to send a packet. When a virtual network is out of credit, the node cannot transmit a packet. As long as there are other active queues, the virtual network is stopped. If, at some point, all queues are empty, a reallocation of credits is assigned to all queues. However, this can lead to a blocking situation, either queues have no credit or queues enjoy credits but no packet to be served. To avoid this problem, the blocked queues can create negative credits and continue to transmit. When the sum of positive and negative credits becomes equal to zero, the system is reset by providing positive credits to all nodes. From this example, available on Xen hypervisor (http://www.xen.org), a queue can be allowed to serve multiple packets as soon as the queue possesses available credits.

13.3.2. *Extensive network virtualization*

Virtualization technology was applied first on routers and this is surely one of the most interesting applications for a virtual machine. However, many other hardware devices can also be virtualized. As an example, we can virtualize WiFi controllers. This is used to manage WiFi access points from a virtual machine placed in a datacenter away from the local area network. The associated access points can have their own protocol stack.

A firewall can also be virtualized. To access a company, the packets are guided to a general virtual firewall which, in turn, may redirect the flow to a virtual firewall dedicated to a specific application. The virtual firewalls can settle all the necessary power to perform the deep packet inspections to detect attacks.

The access points themselves can be virtualized. Simply introduce a hypervisor supporting multiple operating systems. Each virtual access point has its own software management and control. The virtual access points can be independent of each other even if located in the same infrastructure. Similarly, it is possible to virtualize BTS, NodeB, eNodeB, HeNB, etc. This allows us to share antennas between different operators.

Overall, any resource may be virtualized, except sensors, which require a hardware element to measure a value, physical antennas, and senders and receivers. Therefore, a virtual sensor may be defined as a software associated with a physical sensor to perform calculations from the measured values. This software can be changed, hence the name of virtual sensor. The future can be viewed as a virtualization of all physical machines and more explicitly all functions contained in a physical machine. A large number of pieces of network equipment may be virtualized as shown in Figure 13.8.

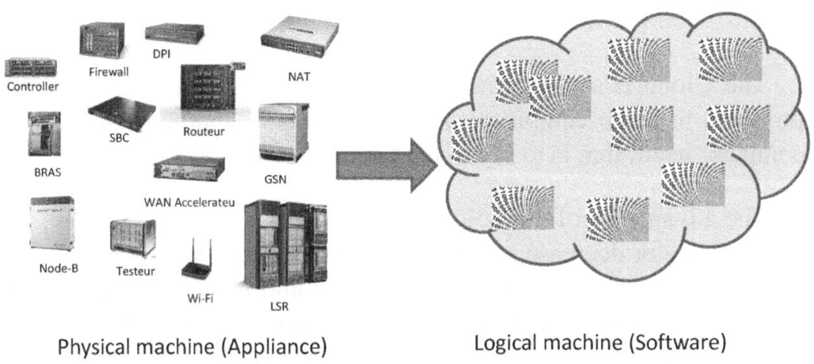

Figure 13.8. *Network equipment virtualization*

In this figure, the operating system supports a variety of protocol stacks adapted to different problems, such as lack of energy, security, Quality of Service, increased availability, etc. Each virtual network can use its own protocol stack and change the protocol stack as soon as necessary, for example when discovering a problem such as a battery almost empty or a DDOS attack.

Virtualizing physical or radio links is also possible. A physical link can be replaced by a virtual link as soon as the client cannot see the difference due to identical performance. WiFi can also be virtualized if the users did not detect they are using a virtual WiFi hotspot. Again the performance of the virtualized system has to be identical to a non-shared infrastructure. For WiFi access points, this will be the case with very high-capacity WiFi technologies such as IEEE 802.11ax.

Finally, virtualization is an important technique, which allows us to introduce new technology painlessly as virtual private networks. Thus, it is possible to test a network running a new architecture without any risk. Virtualization is also a long-term solution to introduce new generation networks like Post-IP proposals. When the network architecture of the future will be perfectly defined, this architecture may come gradually worldwide through virtualization.

13.3.3. *The Cloud*

The Cloud is a paradigm that was introduced around 2005 which proposes to place equipment and software of a company outside the company itself, that is to say in the Cloud.

The advantages of this solution are numerous for the company: this can increase or decrease resources at will. The company is not obliged to oversize the system since it is simple to add new resources as soon as necessary. In addition, the user only pays for what he consumes. The company does not need to employ staff to take over the management of equipment. This is an undeniable advantage for Cloud-based solutions. For the Cloud service provider, the benefits are numerous: sharing of resources between all clients and so underlying server virtualization, systems, software, networks, etc. The statistical multiplexing of resources is pretty good as soon as the number of customers is high.

The Cloud can be fairly well represented by the Internet allowing the connection of the users to datacenters. The Internet is capable of supporting different types of datacenters running business applications. This solution represents an offset of the IT environment of a company to the Internet. The Cloud resources can be extremely diverse: computational resources, storage of information, software, mail service, etc. Depending on the virtualized elements available in the Cloud, several types of Clouds can be defined:

– IaaS (Infrastructure as a Service). This is a first approach, with network resources supported in the Cloud, such as network servers, storage servers and network itself. We can find IP-PBX, firewall, storage servers, etc., and more generally the servers associated with the network infrastructure.

– PaaS (Platform as a Service). This is a second model of Cloud, with, in addition to the infrastructure, the middleware corresponding to the Internet platform. The company's servers only take into account the applications.

– SaaS (Software as a Service). This solution allows the Cloud provider to offer in addition to the infrastructure and the platform, the applications themselves. Almost nothing remains in the company, only Internet terminals. This solution, also known as Cloud Computing, externalizes virtually all servers and corporate networks.

These solutions were strongly developed in 2010s. However, it is quite difficult for a Cloud provider to optimize its own infrastructure and especially the placement of resources available to users. In fact, optimization strongly depends on what is desired. To optimize Cloud operations, performance criteria are important: optimizing performance and power consumption, balancing loads, etc.

The resource location should change over time to achieve energy and cost savings. For example, it could be better to place resources on servers located in an area where it is night on Earth. However, resource migration can be expensive. If optimization concerns the cost of energy, it is clear that resources must be collected on the same machine as close as possible to the company. One can find dozens of different solutions depending on the performance criterion to be optimized.

Virtualization plays a key role as all resources are virtualized resources (server, storage, network equipment, etc.). In particular, connections to the servers in the Cloud must be done through a virtual network set up for this purpose. The virtual network is modified as a function of the virtual routers which are migrating to optimize performance.

One difficulty that arises in a business is to find the best Cloud provider compared to its needs. Unfortunately, finding an email service provider, calculation, storage, specific software, etc., within the same Cloud provider is impossible. The company must ask for multiple cloud providers, complicating the company's decision. To solve this issue, intermediate suppliers appear, which might be called Sky providers. These Sky providers can optimize the selection of the Clouds necessary for a company.

A Sky provider can be seen as the provider of a set of Clouds adapted to a company. Another name for a Sky provider is a Virtual Clouds Provider. The client (the company), instead of searching for all Clouds that will offer him the service he needs himself, addresses the Sky provider capable of managing all access to different Clouds necessary to his business. The largest companies in the late 2010s should be Sky providers.

13.4. Mobile-edge computing

Computing provides a new ecosystem and value chain, and the opportunity for all players within it to collaborate and develop new business models so that they can benefit from each other. Mobile Network Operators (MNOs) can rapidly deploy new services for consumer and enterprise business segments which can help them differentiate their service portfolio. Adding new revenue streams from innovative services delivered to the users can improve the MNO's bottom line while improving end user QoE. New applications whichare aware of the local context in which they operate (RAN conditions, locality, etc.) can open up entire new service categories and enrich the offering to end users. Placing relevant applications on or near the base station not only offers advantages to consumer and enterprise end users, but it also reduces the volume of signaling offloaded to the core network and could also reduce OPEX for the MNOs, compared to hosting in the core.

The MNOs could increase their revenue by charging based on the resource usage (storage, NW bandwidth, CPU, etc.) of each content provider, if such resource usage could be obtained via specific Application Programming Interfaces (APIs) in MEC servers.

Software and application providers can serve the new ecosystem by developing and bringing to the market innovative and ground breaking services and applications that can take advantage of the information on radio network capabilities and conditions available at the base station. The application space is open to anyone: software and application providers, infrastructure vendors and MNOs. The use of open standards and APIs, as well as the use of familiar programming models, relevant tools chain and Software Development Kits are key pillars to encourage and expedite the development of new disruptive applications or the adaptation of existing services and applications to the new Mobile-edge Computing environment. We are going to describe different examples of use cases using Mobile-Edge Computing.

13.4.1. *Use case 1: active device location tracking*

Figure 13.9 shows an example of the active device location tracking use case. This use case enables real-time, network measurement-based tracking of active (GPS independent and network determined) terminal equipment, using "bestin-class" third-party geo-location algorithms within a geo-location application hosted on the MEC server. This provides an efficient and scalable solution with local measurement processing and event-based triggers. It enables location-based services for enterprises and consumers (e.g. on opt-in basis), for example in venues, retail locations and traditional coverage areas where GPS coverage is not available.

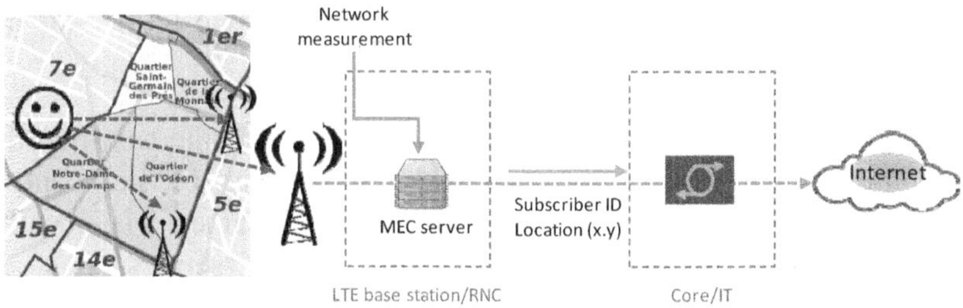

Figure 13.9. *Example of active device location tracking*

13.4.2. *Use case 2: augmented reality content delivery*

An Augmented Reality (AR) application on a smart-phone or tablet overlays augmented reality content onto objects viewed on the device camera (as displayed in Figure 13.10). Applications on the MEC server can provide local object tracking and local AR content caching. The solution minimizes round trip time and maximizes throughput for optimum quality of experience. It can be used to offer consumer or enterprise propositions, such as tourist information, sporting event information, advertisements, etc.

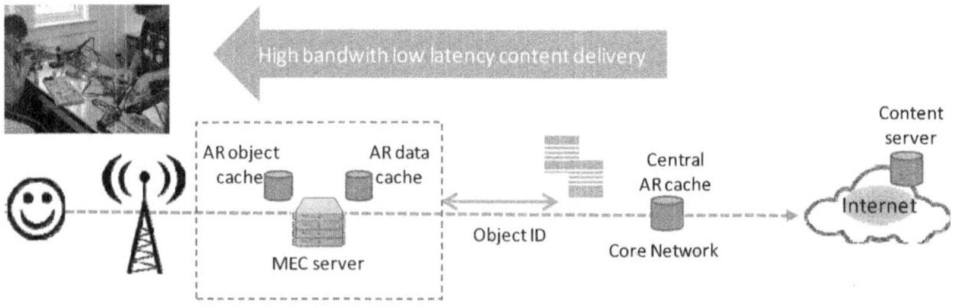

Figure 13.10. *Example of augmented reality content delivery*

13.4.3. *Use case 3: video analytics*

Figure 13.11 shows a distributed video analytics solution which provides an efficient and scalable mobile solution for LTE. The video management application transcodes and stores captured video streams from cameras received on the LTE uplink. The video analytics application processes the video data to detect and notify specific configurable events, e.g. object movement, lost child, abandoned luggage, etc. The application sends low bandwidth video metadata to the central operations and management server for database searches. Applications may range from safety, public security to smart cities.

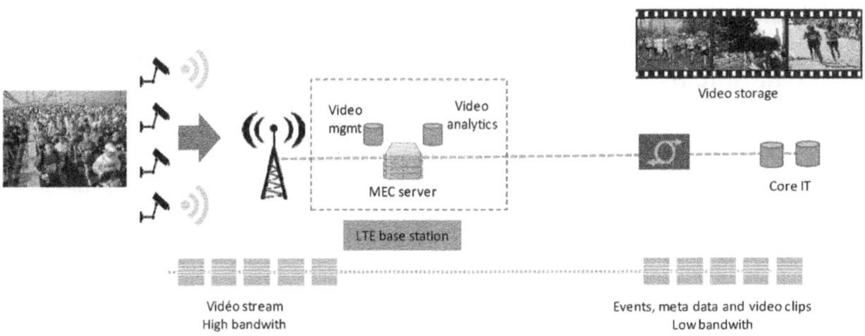

Figure 13.11. *Example of video analytics*

13.4.4. *Use case 4: RAN-aware content optimization*

In this use case, the application exposes accurate cell and subscriber radio interface information (cell load and link quality) to the content optimizer, enabling dynamic content optimization, improving QoE, network efficiency and enabling new service and revenue opportunities. Dynamic content optimization enhances video delivery through reduced stalling, reduced time-to-start and “best” video quality. Figure 13.12 displays an example of RAN-aware content optimization. The concept also enables enhanced use cases, such as promoted content delivery and subscriber throughput boosting.

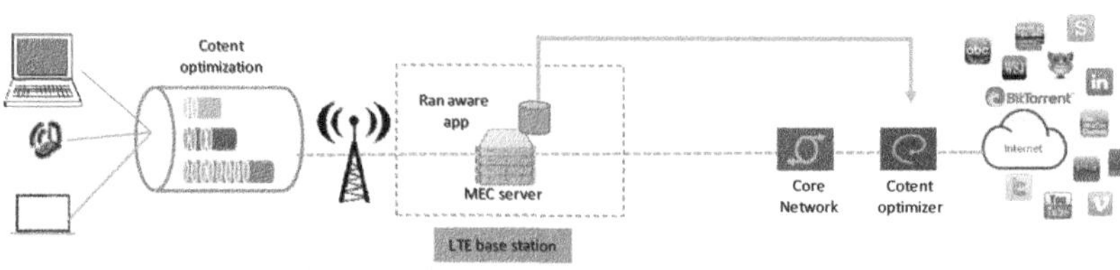

Figure 13.12. *Example of RAN-aware content optimization*

13.4.5. *Use case 5: distributed content and DNS caching*

A distributed caching technology (as shown in Figure 13.13) can provide backhaul and transport savings and improved QoE. Content caching has the potential to reduce backhaul capacity requirements by up to 35%. Local Domain Name System caching can reduce web page download time by 20%.

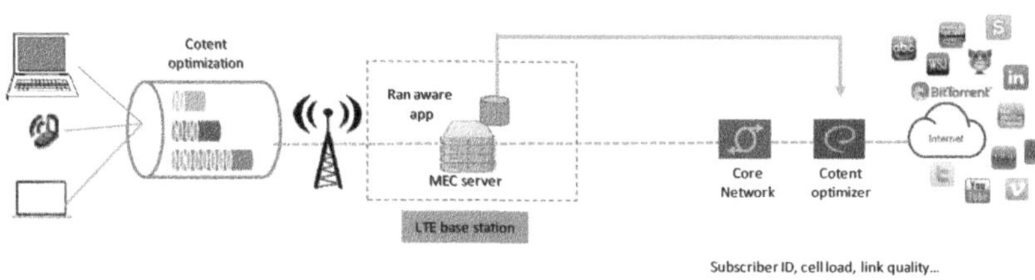

Figure 13.13. *Example of distributed content and DNS caching*

13.4.6. *Use case 6: application-aware performance optimization*

Application-aware cell performance optimization for each device in real time can improve network efficiency and customer experience (see Figure 13.14). It can reduce videostalling and increase browsing throughput. Latency may also be significantly reduced. The solution can also provide independent metrics on application performance (video stalls, browsing throughput and latency) for enhanced network management and reporting.

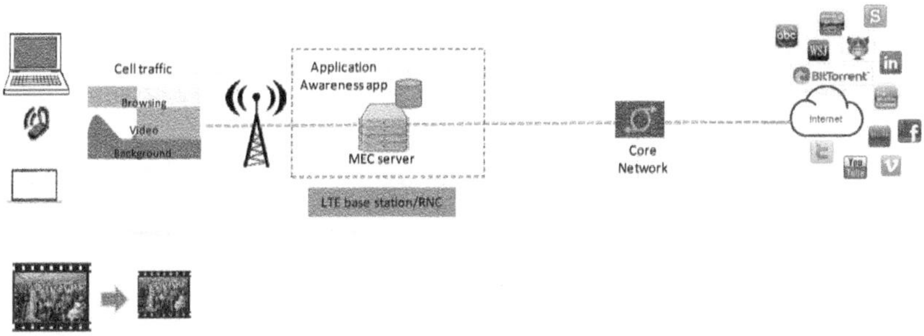

Figure 13.14. *Example of application-aware performance optimization*

13.4.7. *MEC server placement*

The MEC server can be any distance from the terminal equipment. Figure 13.15 provides an overview of various options. The first would be to place the server in the user's network equipment, i.e. in the Home enhanced NodeB. This case is very specific because the MEC server is small and replaces the WiFi broadband router or Home Gateway or Internet Box. The following placement is close to the antenna or base station. Several possibilities are offered to the operators such as next to the eNodeB or in the eNodeB. In the latter case, it is rather the opposite that occurs, the eNodeB is encapsulated into the MEC server. A final possibility which should be widely used by mobile operators is to put a very large MEC server in the RNC for integrating simultaneously the eNodeBs of multiple antennas.

Management of the MEC servers is crucial for mobile operators. Much of the cost is at this level. Figure 13.16 describes an MEC server architecture and management. The server consists of three layers. The first is the hardware that supports the software environment. On this equipment is included a hypervisor that can support virtual machines. These virtual machines bring the various services requested to the MEC server. A first management system corresponds to this layer. This system has to take care of all the hardware infrastructure and virtualization layer.

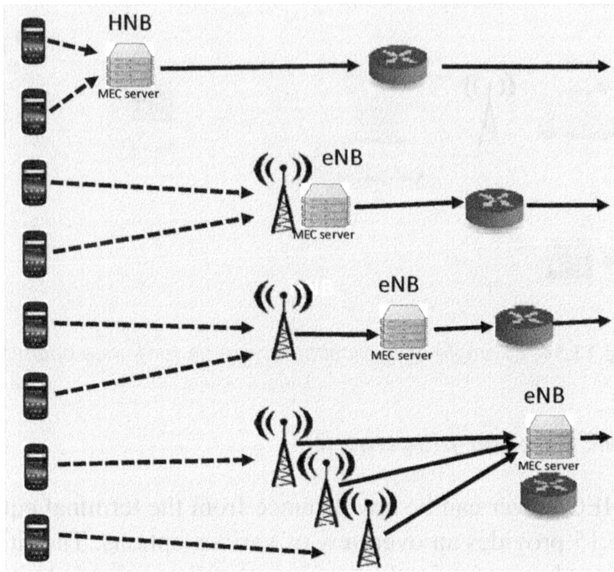

Figure 13.15. *Different MEC server placements*

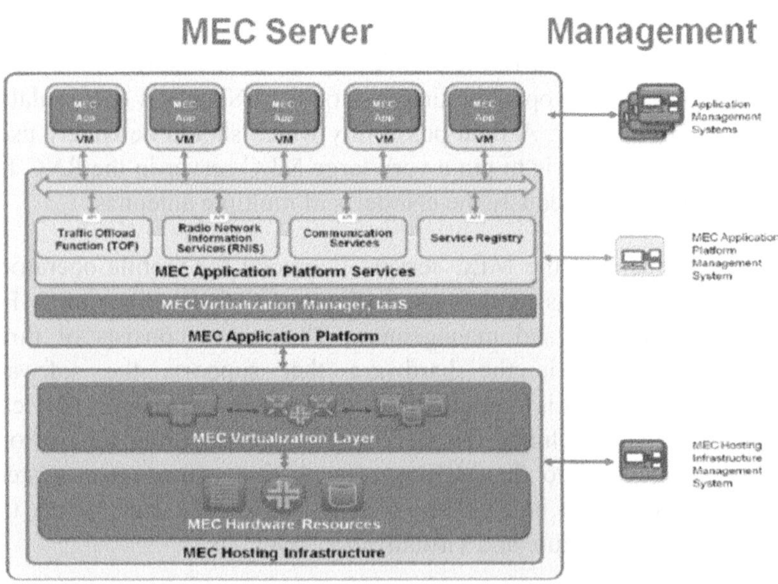

Figure 13.16. *MEC server architecture and management*

The second part concerns the application platform for the MEC server. This platform supports all the functions required for the operation of user communications to the core network or to each other if they are located on the same server MEC. The platform operates RAN control functions, handover schemes, attachment structures, traffic control algorithms, call management and all necessary services to be offered to the customers. This layer of the MEC server is managed by a system that deals exclusively with this platform.

The last part of the MEC server concerns the virtual machines themselves. These virtual machines support the functions described above. These virtual machines can be available permanently, but can also migrate from more central datacenters. Migration of virtual machines is important in power consumption and optimization services. A third system is vested in the management of these virtual machines.

13.5. Conclusion

MEC technologies are subject to strong development and should take much of the mobile network market. In fact, today the discussion focuses on the intelligence provided by the MEC servers. Mobile operators are pushing strongly toward an intelligence coming directly from the ECM server while network manufacturers are rather in favor of a distribution of the intelligence in intermediate equipment located between the terminal equipment and the MEC server. In the first case, it would be rather like the OpenFlow protocol that would be used and in the second case, the I2RS type of protocols could take over.

13.6. Bibliography

[AVA 14] AVRAMOV L., PORTOLANI M., *The Policy Driven Data Center with ACI: Architecture, Concepts, and Methodology*, CISCO Press, 2014.

[AZO 13] AZODOLMOLKY S., *Software Defined Networking with OpenFlow*, Packet Publishing, 2013.

[GEN 14] GENG H., *Data Center Handbook*, Wiley, 2014.

[GOR 14] GORANSSON P., BLACK C., *Software Defined Networks: A Comprehensive Approach*, Morgan Kaufmann, 2014.

[HU 14] HU F., *Network Innovation through OpenFlow and SDN: Principles and Design*, CRC Press, 2014.

[MOR 14] MORREALE P.A., ANERSON J.M., *Software Defined Networking: Design and Deployment*, CRC Press, 2014.

[NAD 13] NADEAU T.D., GRAY K., *SDN: Software Defined Networks*, O'Reilly, 2013.

[RHO 14] RHOTON J., DE CLERCQ J., NOVAK F., *OpenStack Cloud Computing: Architecture Guide*, RP Publisher, 2014.

[SAB 13] SABOOWALA H., ABID M., MODALI S., *Designing Networks and Services for the Cloud: Delivering Business-Grade Cloud Applications and Services*, CISCO Press, 2013.

[SHU 13] SHUKLA V., *Introduction to Software Defined Networking – OpenFlow & VxLAN*, Createspace, 2013.

[WEN 13] WEN H., TIWARY P.K., LE-NGOC T., *Wireless Virtualization*, Springer, 2013.

[YEL 14] YELURI R., CASTRO-LEON E., *Building the Infrastructure for Cloud Security: A Solutions View*, Apress Open, 2014.

Conclusion

Wireless and mobile networks are always changing. The move to 5G will allow us to be in a world where everything is connected, always, everywhere. The flow of each client continues to grow steadily with new releases. The share of wireless connections will increase to 85% in 2020 counting the throughput of all the connections in the world from a terminal equipment. Wireless takes place inexorably of the wire. In 2020, we can consider that the wired access is disappearing and all users, objects, sensors, things are connected wirelessly.

However, there are still problems to solve and they are relatively numerous. First, there is the problem for all human or machine connections which can cause power and saturation problems in relation to the IP world that generally requires rather a lot of energy. This means that new protocols appear that consume much less energy but require gateways to access the Internet. Second, everywhere there could be a problem, because if it is relatively simple to be connected in cities, connections are much more limited inside the buildings and in sparsely populated areas. Third, being connected always means a network availability with a very low failure rate and therefore duplications of infrastructure that greatly increase the global cost.

Finally, the impact of Cloud-based networking is modifying wireless networks since the user data and applications are stored in datacenters. The changes are going in four directions:

– user data and applications are located in central datacenters. Over-the-top (OTT) content concerning delivery of audio, video and other media over the Internet is the main paradigm to be used;

– telecommunication operators are moving toward centralized Mobile Edge Computing MEC servers, i.e. toward a distribution of data and applications in smaller datacenters closer to users but with a centralization on these small datacenters;

– network manufacturers prefer to push intelligence and control schemes into their edge equipment, i.e. distributing data and applications in edge equipment;

– finally, users see their terminal equipment becoming more powerful in order to ensure for themselves the control of their own data and applications.

In summary, research continues and the struggle concerns the location of the control algorithms. The definition of future wireless network architectures is just beginning.

Index

E

F

N, O

P

Q, R

S

V

W

X, Z

Other titles from

iSTE

in

Networks and Telecommunications

2016

BENSLAMA Malek, BATATIA Hadj, MESSAI Abderraouf
Transitions from Digital Communications to Quantum Communications: Concepts and Prospects

CÂMARA Daniel, NIKAEIN Navid (Eds.)
Wireless Public Safety Networks 2: A Systemic Approach

CHIASSERINI Carla Fabiana, GRIBAUDO Marco, MANINI Daniele
Analytical Modeling of Wireless Communication Systems (Stochastic Models in Computer Science and Telecommunication Networks Set – Volume 1)

EL FALLAH SEGHROUCHNI Amal, ISHIKAWA Fuyuki, HÉRAULT Laurent, TOKUDA Hideyuki
Enablers for Smart Cities

PEREZ André
VoLTE and ViLTE

RAMOS Angel, LAZARO Antonio, GIRBAU David, VILLARINO Ramon
RFID and Wireless Sensors using Ultra-Wideband Technology

SENOUCI Mustapha Reda, MELLOUK Abdelhamid
Deploying Wireless Sensor Networks

2015

BENSLAMA Malek, KIAMOUCHE Wassila, BATATIA Hadj
Connections Management Strategies in Satellite Cellular Networks

BENSLAMA Malek, BATATIA Hadj, BOUCENNA Mohamed Lamine
Ad Hoc Networks Telecommunications and Game Theory

BERTHOU Pascal, BAUDOIN Cédric, GAYRAUD Thierry, GINESTE Matthieu
Satellite and Terrestrial Hybrid Networks

CÂMARA Daniel
Bio-inspired Networking

CÂMARA Daniel, NIKAEIN Navid (Eds.)
Wireless Public Safety Networks 1: Overview and Challenges

CUADRA-SANCHEZ Antonio, ARACIL Javier
Traffic Anomaly Detection

JARRY Pierre, BENEAT Jacques N.
Digital Communications

LE RUYET Didier, PISCHELLA Mylène
Digital Communications 1: Source and Channel Coding

PEREZ André
LTE and LTE Advanced: 4G Network Radio Interface

PISCHELLA Mylène, LE RUYET Didier
Digital Communications 2: Digital Modulations

PUJOLLE Guy
Software Networks

2014

ANJUM Bushra, PERROS Harry
Bandwidth Allocation for Video under Quality of Service Constraints

BATTU Daniel
New Telecom Networks: Enterprises and Security

BEN MAHMOUD Mohamed Slim, GUERBER Christophe, LARRIEU Nicolas, PIROVANO Alain, RADZIK José
Aeronautical Air–Ground Data Link Communications

BITAM Salim, MELLOUK Abdelhamid
Bio-inspired Routing Protocols for Vehicular Ad-Hoc Networks

CAMPISTA Miguel Elias Mitre, RUBINSTEIN Marcelo Gonçalves
Advanced Routing Protocols for Wireless Networks

CHETTO Maryline
Real-time Systems Scheduling 1: Fundamentals
Real-time Systems Scheduling 2: Focuses

EXPOSITO Ernesto, DIOP Codé
Smart SOA Platforms in Cloud Computing Architectures

MELLOUK Abdelhamid, CUADRA-SANCHEZ Antonio
Quality of Experience Engineering for Customer Added Value Services

OTEAFY Sharief M.A., HASSANEIN Hossam S.
Dynamic Wireless Sensor Networks

PEREZ André
Network Security

PERRET Etienne
Radio Frequency Identification and Sensors: From RFID to Chipless RFID

REMY Jean-Gabriel, LETAMENDIA Charlotte
LTE Standards
LTE Services

TANWIR Savera, PERROS Harry
VBR Video Traffic Models

VAN METER Rodney
Quantum Networking

XIONG Kaiqi
Resource Optimization and Security for Cloud Services

2013

Assing Dominique, Calé Stéphane
Mobile Access Safety: Beyond BYOD

Ben Mahmoud Mohamed Slim, Larrieu Nicolas, Pirovano Alain
Risk Propagation Assessment for Network Security: Application to Airport Communication Network Design

Beylot André-Luc, Labiod Houda
Vehicular Networks: Models and Algorithms

Brito Gabriel M., Velloso Pedro Braconnot, Moraes Igor M.
Information-Centric Networks: A New Paradigm for the Internet

Bertin Emmanuel, Crespi Noël
Architecture and Governance for Communication Services

Deuff Dominique, Cosquer Mathilde
User-Centered Agile Method

Duarte Otto Carlos, Pujolle Guy
Virtual Networks: Pluralistic Approach for the Next Generation of Internet

Fowler Scott A., Mellouk Abdelhamid, Yamada Naomi
LTE-Advanced DRX Mechanism for Power Saving

Jobert Sébastien *et al.*
Synchronous Ethernet and IEEE 1588 in Telecoms: Next Generation Synchronization Networks

Mellouk Abdelhamid, Hoceini Said, Tran Hai Anh
Quality-of-Experience for Multimedia: Application to Content Delivery Network Architecture

Nait-Sidi-Moh Ahmed, Bakhouya Mohamed, Gaber Jaafar, Wack Maxime
Geopositioning and Mobility

Perez André
Voice over LTE: EPS and IMS Networks

2012

AL AGHA Khaldoun
Network Coding

BOUCHET Olivier
Wireless Optical Communications

DECREUSEFOND Laurent, MOYAL Pascal
Stochastic Modeling and Analysis of Telecoms Networks

DUFOUR Jean-Yves
Intelligent Video Surveillance Systems

EXPOSITO Ernesto
Advanced Transport Protocols: Designing the Next Generation

JUMIRA Oswald, ZEADALLY Sherali
Energy Efficiency in Wireless Networks

KRIEF Francine
Green Networking

PEREZ André
Mobile Networks Architecture

2011

BONALD Thomas, FEUILLET Mathieu
Network Performance Analysis

CARBOU Romain, DIAZ Michel, EXPOSITO Ernesto, ROMAN Rodrigo
Digital Home Networking

CHABANNE Hervé, URIEN Pascal, SUSINI Jean-Ferdinand
RFID and the Internet of Things

GARDUNO David, DIAZ Michel
Communicating Systems with UML 2: Modeling and Analysis of Network Protocols

LAHEURTE Jean-Marc
Compact Antennas for Wireless Communications and Terminals: Theory and Design

RÉMY Jean-Gabriel, LETAMENDIA Charlotte
Home Area Networks and IPTV

PALICOT Jacques
Radio Engineering: From Software Radio to Cognitive Radio

PEREZ André
IP, Ethernet and MPLS Networks: Resource and Fault Management

TOUTAIN Laurent, MINABURO Ana
Local Networks and the Internet: From Protocols to Interconnection

2010

CHAOUCHI Hakima
The Internet of Things

FRIKHA Mounir
Ad Hoc Networks: Routing, QoS and Optimization

KRIEF Francine
Communicating Embedded Systems / Network Applications

2009

CHAOUCHI Hakima, MAKNAVICIUS Maryline
Wireless and Mobile Network Security

VIVIER Emmanuelle
Radio Resources Management in WiMAX

2008

CHADUC Jean-Marc, POGOREL Gérard
The Radio Spectrum

GAÏTI Dominique
Autonomic Networks

LABIOD Houda
Wireless Ad Hoc and Sensor Networks

LECOY Pierre
Fiber-optic Communications

MELLOUK Abdelhamid
End-to-End Quality of Service Engineering in Next Generation Heterogeneous Networks

PAGANI Pascal *et al.*
Ultra-wideband Radio Propagation Channel

2007

BENSLIMANE Abderrahim
Multimedia Multicast on the Internet

PUJOLLE Guy
Management, Control and Evolution of IP Networks

SANCHEZ Javier, THIOUNE Mamadou
UMTS

VIVIER Guillaume
Reconfigurable Mobile Radio Systems

Printed and bound by CPI Group (UK) Ltd, Croydon, CR0 4YY
13/09/2026
14961079-0004